Conductive Electronic Weapons

And Their Faults

Trevor Langevin

HUMBER LIBRARIES LAKESHORE CAMPUS
3199 Lakeshore Blvd West
TORONTO, ON.   M8V 1K8

Copyright © 2011 by Trevor Langevin

First Published: 2011

All rights reserved. Without limiting the rights under copyrights reserved above, no part of this publication may be reproduced, stored in or introduced into a retrieval system, or transmitted, in any form or by any means (electronic, mechanical, photocopying, recording or otherwise), without the prior written permission of the copyright owner. Book reviewers and news media such as Newspapers, Television & Radio may quote short excerpts.

Copyrights & Trademarks remain the property of their respective owners. In no event will the author or publisher be held liable for any use or misuse of information, or Copyright & Trademark errors contained within this book.

Formatted using Scribus

Printed and bound in the United States of America
by Lulu, Inc.
www.Lulu.com

ISBN: 978-1-257-09734-0

Table of Contents

Introduction   1

CEW Report
- The differences between subject testing and real world   5
- Electrical discharge complications to the body   7
- Effects of high voltage discharge on water   8
- The heart or cardiac muscle   12
- Tetanus and Clonus   14
- Ventricular Tachycardia and Ventricular Fibrillation   15
- Cardiac pulse rhythm   15
- NMI Neuromuscular Incapacitation   16
- Taser operation   16
- Nerve muscles and stimulation   17
- Troponin   18
- Defibrillator's   18
- Power output of CEW's   20
- Electrostatic discharge   22
- Inductance of the wire   26
- Static electricity over a carpeted surface   27
- Influencing the output   30
- Excited Delirium   35
- Alcohol, cocaine, Nicotine and Amphetamines   39
- Fight or Flight response   40
- Anaesthesia, Sedation   42
- Long term side effects of electrocution   48

Chapter 2 Updates April 12, 2008
- Stinger Systems supplies information   49
- Flat batteries can affect Taser records   50

Chapter 3 April 21, 2008
- Question and answers from reviewers   51

Chapter 4 Updates August 17, 2008
- Updates from the Inquiry   59
- Dr. Sloane and Dr. Tseng review.   59
- Cartridge wires shorting out   66

Table of Contents

Chapter 5 Updates January 15, 2009

Venous gases	67
Taser International and the U.S. Air Force	71
Taser International disputes the CBC testing	77
Taser International replies to the CBC testing	80
Cameron Ward on the internal RCMP investigation	82
News reports	83
How far did the RCMP have to drive	88
YVR Remodels the arrivals area	90

Chapter 6 Updates April 30, 2009

Study raises concerns over Tasers' safety	91
Dziekanski post mortem review	98

Chapter 7 Updates July 27, 2009

Dziekanski toxicology review	109
What is Lactic Acidosis?	110
What is Rigor Mortis?	111
Review of the probe impact locations	112

Chapter 8 Updates April 4, 2010

Taser International inc. sues Canadian Government	117
Avoid aiming taser at chest: Manufacturer to cops	118
Taser issues new warning on shooting into chest area	121
Police hedge on Taser use after new rules	124
More details of commissioner of complaints report,,,	125
Mounties too quick to Taser Dziekanski: report	134
Taser timeline in Canada	136
First test of the new Taser X3	138

Chapter 9 Fallout December 10, 2010

Mountie who tasered Dziekanski sues CBC	139
RCMP tarnished by Dziekanski case: expert	140
RCMP apology falls short	141
Braidwood's final report in gov't hands	143
Braidwood finds officer use of taser not justified,,,	144
Taser Int'l calls for Braidwood report to be overturned	150
Taser claims it has lost business over inquiry,,,	153

Table of Contents

Taser maker challenges Braidwood report 153
Braidwood inquiry taser findings upheld by court 155
Tasers's challenge to overturn Braidwood Inquiry dismissed 156
Taser Int'l wins patent infringement against Stinger Systems 158
The CBC / Radio Canada test update 159

Chapter 10 Conclusions & Recommendations
Behavioral analysis, the unanswered questions 163
Behavioral analysis of the RCMP officers 164
Behavioral analysis of the RCMP management 164
Behavioral analysis of the manufacturer 164
1) Modifications recommended by the CEW report 165
2) Taser International discards the testing results 167
3) Electrical energy can cause tachycardia 168
4) What other recommendations have been proposed 169
5) An avenue for future research 171
6) What this hypothesis covers 171
7) What happened to Dziekanski 172
8) Suitability for commercial aircraft use 174
9) The United Nations Committee Against Torture (CAT) 177
10) What the stun gun has become 179

Chapter 11 Braidwood Inquiry Medical Findings
The Braidwood Inquiry reports 181
The eight findings from the inquiry 182
Excited Delirium 184

Chapter 12 Updates April 5, 2011
Update: The AECP recognizes some cases of ExDS 189
Update: The CMAJ finds Taser caused seizure. 190
Datrend Systems Verus One; ECD / CEW analyzer. 194

Appendix A: Static Electron Detector 195
Appendix B: CEW Report & Updates Sources 197
Appendix C: Braidwood Inquiry List of Presenters 205
Appendix D: List of the Dead 209
Appendix E: Comparison of Devices 227
Notes: 231

*

A simple memorial to the lives that have been ruined.

For all those who have lost a loved one,
or to those who no one remembers,
or to the officers who don't understand.

Rest in Peace

*

Introduction

I work for a major airline in Canada, on the Health and Safety Committee. This unofficial report, now a book was not intended to investigate the death of Mr. Robert Dziekanski. It has been rewritten to include it as the course of the investigation progressed. This report was originally just a few lines of notes, which greatly expanded. Its purpose was to investigate if a probe launching designed CEW would be a viable alternative to a firearm, as carried by police and sky marshals on commercial aircraft.

The Dziekanski incident happened not long after this document was started, and so those results have been included into this report. It was never intended to be a scientific study with perfectly referenced sentences, but the amount of information collected was overwhelming at times.

This report focuses on the current line of probe launching CEW's:
The M34000™ the M26™ the X26™ and S-200™ Newer devices and manufacturers should be tested for the same faults also.

Also key to the investigation is the unexplained acidosis of the blood, as not explained in the testing literature of the Conducted Electronic Devices, and the BCOPCC_final report. With the progression of information, it has been realized that some or all of the results found could happen every time a CEW is used. This report uses a Human Factors standpoint, instead of just a medical or technical standpoint.

The author also believes that there is a fault in the design of the Conducted Electrical Weapons as they are currently manufactured, and a flaw in the testing procedures, and it will be described. Both of these problems have resulted in the current death count (When submitted to the inquiry) of 336 in North America, now well over 500. Not every person who gets shot by a CEW is a criminal.

The CEW's were never intended to take lives, but it has left the officers feeling as unintentional executioners, wondering when it will be their turn. The intent of the device is to save lives by not having to use a gun. This goal has not entirely been met.

It is the intention of this report to source the faults and find a reasonable solution to the problems. The problems specifically are that people are dying within a short time frame (from seconds to days) after being shot by a CEW. It is not the intent of this report to remove a tool from the police or military arsenal. That will be a consensus of the scientific community, and the community at large to make that decision.

In Canada, the CEW is a prohibited weapon, and the only purpose of the device is to cause harm onto another person. As such a device was not attainable for testing. The Richmond British Columbia RCMP detachment after repeated requests has provided zero assistance with the making of this report. This includes contact in person, by email, and not returning phone call messages, or pulling a fired cartridge out of the garbage. Several officers did try to pursue the issue on my behalf, but were stopped by the local training person, whom has since left the RCMP and shall remain nameless!

Taser International denied providing information not contained in the products specification data sheets to the author, stating that all information was proprietary. This was by telephone and email's. After sending copies of this report and the updates to Taser International, they refused to acknowledge all further contact. The only two questions I ever had for them was what was the AWG wire size, and how were the electrical wires connected in the probes. Pictures provided by others would eventually answer that question.

Stinger Systems has provided a report on recent testing of pigs, and was very good at returning all email questions. However due to the litigation with Taser International, details of the documents were not going to be included in this report. It was reviewed or correlation of data and the outcome was that it did not conflict with this hypothesis. Stinger later responded after the report was sent for review that the details could be included.

Introduction · 3

The report by the British Columbia Office of the Police Complaint Commissioner "Taser Technology Review final Report" (BCOPCC) was the main source of the medical information to be investigated, that this report relies on.

This report is not final. Other medical reactions and symptoms not listed are very possible. The author is extremely grateful for all the companies, and individuals who's web sites, email and telephone contacts contributed to the making, and reviewing of this report.

The book has been slightly reorganized and expanded from the reports submitted to the Braidwood Inquiry, taking in updated information and corrections since the trial. The first chapter is the original report submitted to the first phase of the Braidwood Inquiry. It incorporates the human factors look at the stun guns and how they can be faulted. The technical details surrounding the weapons use on Mr. Dziekanski is also looked at. Chapters 2-8 cover the updates since the start of the inquiry, and responses to events that happened during the course of the inquiry. The intent is not to cover the story of the trial, but document the information surrounding the stun gun, the manufacturer, and any links to the postmortem. Chapter 9 deals with the after events or results of the inquiry, and chapter 10 is a combination of conclusions and updates since submission. Chapter 11 is the medical findings from the inquiry, and Chapter 12 is last minute updates found while formatting into the book.

What is Human Factors? Simply, it is the multi directional field of how the Human interacts with its surroundings. Psychology, industrial design, engineering, operations research, and anthropometry. By limiting the research to the medical and electrical fields has stagnated the search for answers into why people are dying after applications from these devices.

The author recognizes that other medical, scientific and engineering opinions exist, and he welcomes the feedback. However he asks that if any verbal diarrhea is included in the correspondence, that it be used in a constructive and positive manner to keep any important information from being unduly edited in any follow up revisions.

Stinger Systems was not brought into the Braidwood Inquiry because at that time their product had not been approved for use in Canada.

This book was written between 2007 to 2011. As usual with internet sources, the web sites that are here today, are sometimes gone tomorrow. No information in this book was manufactured. If there was no confirmed source, then it was not included. Information sources were acquired world wide, and no attempt was made to recompile or correct for their spelling.

Much information came from other media sources. The author choose to include news articles on the subject verbatim. This does give the book an appearance that the author took the easy road and just combined a bunch of footnotes, however the articles are right to the point, and the meaning doesn't translate when recompiled into more creative literal works.

The author would like give a special thanks for two web sources that have tirelessly kept track of nearly all the issues and developments with the stun guns and their manufacturer.

http://Truthnottasers.blogspot.com
http://www.excited-delirium.blogspot.com

Together they have documented nearly everything that the manufacturer has done, and changes implemented.

Amnesty International was contacted to find out why their count of the dead is less than the TNT web site. Their reply was that they do not have a person dedicated to seeking this information out, they rely on individuals sending the information to them. The TNT web site owner on the other hand is actively searching the news links daily for any and all information related to stun guns, and their manufacturer.

Unsolicited Report On:
Conductive Electronic Weapons

The differences between subject testing and real world:

In documented testing, and videos found on the Internet, it has been noticed that subjects that volunteer to be stunned are restrained, and have not been subjected to recent physical activity, such as running around the block to get the circulation going. Also not all training subjects get darts penetrating into the muscle, but alligator clips or a metallic rod connected by insulated wires to the CEW that don't break the skins barrier. In those cases, the device appears to work as intended. The Alligator clips are attached to clothing and targeted into large muscle groups such as the upper back, shoulders or buttocks. The metallic rod is held in the hand, with another grounding wire connected by alligator clip on the leg.

When medical subjects are being tested such as the pigs or the dogs, they are under sedation. (65) Any animal or person under sedation has a reduced pain threshold, and reaction time. There is therefore not a requirement for the bodies to be suppling a lot of blood flow to the muscles. A higher resistance is therefore encountered. There is also not the visual stimuli of a dangerous situation when a person or animal is sedated, so there is no physical changes to contend with such as super human strength and the downward side effects that go with it.

Medical equipment is designed to protect against static electricity, and different requirements exist for their manufacture. A laboratory setting can be free of the differences that happen in the environment.

It has come to the author's attention that some of the researchers who have published reports stating that no ventricular fibrillation was found did not use the off-the-shelf model of CEW, as reported by another researcher.

Taser International built a special machine to mimic the output signal for testing. It is capable of a variable power output, and on the low setting did not find fibrillation, but did on the high settings. This of course opens up several avenues of speculation. The other researchers who found fibrillation, used off-the-shelf devices. (28)

In actual field use there is of course little or no control of environmental factors or where the darts will actually penetrate. A worst case location is one in the Vein and the other in an Artery, but the author can not find testing on that? Subjects are usually fleeing from the police, agitated, aggressive, heightened adrenaline levels, all of which would have excited circulation levels. The higher the blood flow, the more oxygenated the areas become. A lower resistance is then encountered, and a greater pulse of energy delivered to the body. The document "Electrical Evaluation of the Taser M-26 Stun Weapon, Final Report" covers several power output levels from a laboratory setup. The graphs on page 25 are examples of power spike problems encountered with the spark gap design.

CBC.CA / National (28) in Canada reported on January 30, 2008 reported that independent testing on pigs to confirm testing done by Taser International (1 pig in 1996 and 5 dogs in 1999) and possibly others, that heart arrhythmia did happen and that some of the test subjects died from fibrillation. The animals chests was opened and visually inspected. *Stinger Systems was not mentioned.* Another written report from Taser International shows 2 pigs being tested by the X26™, at an unknown date. (11,26)

Wayne McDaniel is in charge of the team at the University of Missouri that did the testing for both Taser International and Stinger Systems on the pigs and dogs. He was consulted and provided some supporting documents. He also assisted with some of the review for the original report. (26)

Dr. Robert Walter who has also done animal studies and found ventricular fibrillation from CEW's was also contacted and provided information for the original report. (27,141)

Electrical discharge complications to the body:

The assumed method of operation.

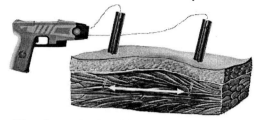

Simply put, the high voltage electrical discharge is only suppose to effect the nervous system and muscle around the darts. Electrical current is only suppose to flow only between the two darts. No other side effects are suppose to happen. (31, 24)

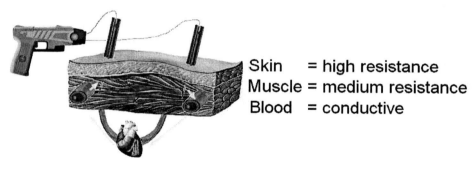

Skin = high resistance
Muscle = medium resistance
Blood = conductive

In reality, where the darts land in the muscle mass is irrelevant. Electricity will always choose to follow the path of least resistance before a path of higher resistance! Simply put, skin has a very high resistance, muscle has a medium resistance, and blood is conductive. The muscle and skin resistance can be lowered proportionally by the amount of blood that has irrigated the area. Electrolytes in the blood are the key ingredient that allows for conductivity. (74,5,40)

For example: When a person is sitting around, or not doing much movement or exercise, there is not a requirement for the body to supply the muscles with a lot of blood flow. The electrical resistance of the muscles would therefore be high. When a person has been exercising and sweating (or running from the police), the muscles are irrigated with blood, causing the electrical resistance to be lowered. This would allow electricity to enter the body easier, travel further and with higher currents.

In cases where the electrical device is not a projectile dart, the addition of sweat or water on the surface of the skin could also produce the same results, under similar conditions. It would depend on the conductivity of skin to transfer the electrical charge through the capillary network, into the blood, in a longer time frame.

The Acidosis of the blood will be covered first. It specifically is located in the Blood Plasma (68), which is over 92 percent water. (42)

Effects of high voltage discharge on water:

Oxygen = O_2 (105,37)
Ozone = O_3 (71,37)
Water = H_2O (106)
Carbonic Acid = H_2CO_3 (72,138,139)
Carbon Dioxide = CO_2 (107)

1) Electrolysis of water $2H_2O \rightarrow H_2 + O_2$
2) Urea and Ozone $(NH_2)2CO + O_3 \rightarrow N_2 + CO_2 + 2H_2O$
3) Carbonic Acid $CO_2 + H_2O \rightleftharpoons HCO_3^- + H^+$

Note: This three stage chemical reaction on these pages is the authors hypothesis for Acidosis after an electrical discharge (1).

Water can be split into its constituent elements, Hydrogen and Oxygen, by passing an electrical charge through it, AC or DC. This process is called Electrolysis. The reaction is greater with DC current, but also works with AC. The electricity causes a chemical reaction to start, and when passed through a body causes the Hydrogen in the blood plasma's water to separate from the Oxygen molecules. The Oxygen molecules want to recombine, the easiest is the other free Oxygen molecules, to become Ozone (O_3). Ozone is a limited time element, and will decompose back into Oxygen (O_2 and a free Oxygen molecule) usually in no more than 30 minutes. The molecules then become free again to recombine.

Ozone has harmful effects on the respiratory system and body.

When Electrolysis starts this reaction, the Plasma effected essentially becomes unusable by the body. Another reaction from Ozone is that it completely decomposes Urea. When CO_2 enters the blood from various cells, it is combined with water to produce Carbonic Acid. It then has a H⁺ taken away from it to become Bicarbonate (HCO_3^-). In order to transport the Bicarbonate that is in the blood stream out of the body, it enters another Red Blood Cell, has H⁺ attached to it to form Carbonic Acid once again, then has H_2O taken away from it and is expelled from the red blood cell as CO_2. Then the Carbon Dioxide is permitted to be expelled out of Capillaries and into the lungs. Hemoglobin, the protein in the Red Blood Cell that carries the Oxygen, will displace an Oxygen molecule for a CO_2 molecule, this known as the Haldane effect. (79)

Carbonic Acid only ever exists in solutions in equilibrium with Carbon Dioxide, and so the concentration of H_2CO^3 is much lower than the concentration of CO_2. Reversed it becomes $H_2CO_3 \rightarrow CO_2 + H_2O$. The equilibrium between Carbon Dioxide and Carbonic Acid is very important for controlling the acidity of body fluids, and almost all living organisms have an enzyme, Carbonic Anhydrase, which catalyses the conversion between the two compounds, increasing the reaction rate by a factor of nearly a billion. This rate would fluctuate with the electrolyte levels.

Rapid fluctuations of the Blood, Urea, Nitrogen (BUN) (73) areas of the body directly effects Electrolyte levels. A dangerous condition known as Hyponatremia (water intoxication) (111) would develop, if the body receives too much water too quickly to compensate for the imbalance. ADH levels (Anti Diuretic Hormone) (75) caused by the Ozone Urea reaction would add complications. Acute Renal Failure (76) from Oliguria and Anuria (Urea production problems) is one. The Nitrogen (77) and other gas components suddenly reappearing would also lead to an Asphyxiation hazard and possibly, Decompression Illness. (88) The body tries to keep the pH (potential of Hydrogen) (78,20) level to 7.4. When the pH level drops below 7.4, it becomes a condition called Acidosis. Below 7.2 cardiovascular complications occur, and below 7 is eventual death, but not from this reaction alone. (20,21,22)

The condition of Acidosis has been documented in testing of the CEW devices (2), and also in Autopsy reports from people subjected to the devices, but the reasons for its presence has never been explained. The only way for the body to handle acidity is to neutralize it with water, bicarbonate or respiration. When there is an acidic rise within the body, the body usually has a small store of water available to deal with it. For example; When you exercise, and start to sweat, and that's the store of water the body is using. The faster you loose it, the more you have that need to get a drink to replenish it. If there is not enough water, then Acidity will continue to rise, causing the pH levels to drop. The bodies organs will take the full force of the toxins until neutralization or death from complications of the pH imbalance. Urine is another method the body uses to removes toxins. The Urea Ozone reaction would hinder that body function, leaving the toxins in the blood stream. Nitrous Oxide (N_2O) occurs in the body to relax muscles, would see a significant increase from returning Oxygen and Nitrogen. (110)

Other chemical reactions from Nitrogen when combined with returning Oxygen, Nitric Oxide (NO) and Nitrogen Dioxide (NO_2). Reperfusion Injury (80) is also a possible long term side effect. Electrolytes in the blood that are widely fluctuating such as Sodium, controls electrical signals within the bodies nervous system and Potassium imbalances cause heart arrhythmias. Chloride levels maintain the bodies fluid levels. Increases or decreases cause deleterious and fatal effects. Bicarbonates are what is monitored to measure blood pH, and Acidity is based on the concentrations of Carbon Dioxide in the blood.

Heart

This is a diagram of a simplified circulatory system. In the human body it takes about 1 minute to circulate the blood supply. If an electrical charge was initiated to ¼ of the system, the blood plasma in that area would become unusable to the body. The only way for the body to cleanse itself of toxic blood is to have the heart pump faster. For every 1 minute, the heart is encountering the toxins for 25 seconds.

As the blood continues to circulate, it will eventually disperse from being one solid area of toxins, to be roughly evenly distributed throughout the body, before neutralization. The key reaction time of this event would be the first 30 minute. This is keeping with the known time frame properties of Ozone. After that the blood would slowly loose its acidity until the chemical interactions are complete. This time limited chemical reaction would be true in a body even after death. Autopsies are typically not done at the time of death to verify this, but later on. The acidosis in the testing reports has a time frame of 30 minute increase and 30 minutes decrease. A Oxygen / Hydrogen reaction should also cause minor heating. (112)

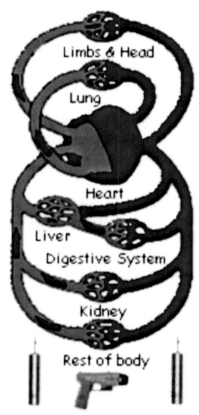

In a little more detailed diagram, you get an idea of how complicated the reaction can get depending on where the electricity is distributed. In cases of multiple electronic distributions through the body, it is very possible that all of, or the majority of the blood can be effected. In that case, the body just can't react to that situation. The body breaths in Oxygen, but the blood would be contaminated and not absorbing the incoming Oxygen. Carbon Dioxide levels would increase as cardiac difficulties increase. Lactate (143) is produced by muscles in a time of decreased oxygen during the bodies Alanine Cycle. A byproduct of this is Urea, which is completely decomposed by Ozone. Troponin is a marker that heart muscle damage has occurred, either electrical or chemical. Obviously lack of Oxygen and Water would effect all the bodies functions.

A graph has been produced to show the stages of the electrolysis reaction on page 57.

The heart, or cardiac muscle:

While the heart is a unique muscle within the body, it is a muscle and has the same weaknesses as other muscles in the body, such as strain and overexertion. The faster it pumps the more it's exerting itself. Just as in any piece of racing machinery, if a movement out of sequence happens, then something will get strained or broken. If the individual's heart is not conditioned to run at accelerated speeds, sudden increases or decreases, it will get strained. Unlike other muscles, it can't stop to recuperate. All it takes is one uncoordinated movement to get it into a fibrillation condition. (36,81,83,84,89)

To create an action potential (a muscle movement) all that is required is for the voltage to go above a certain point. (82) The interior of a resting muscle fiber has a resting potential of about −95mV. The influx of sodium ions reduces the charge, creating an end plate potential. If the end plate potential reaches the threshold voltage (approximately −50mV), sodium ions flow in with a rush and an action potential is created in the fiber. The action potential sweeps down the length of the fiber just as it does in an axon.

No visible change occurs in the muscle fiber during (and immediately following) the action potential. This period, called the latent period, lasts from 3–10msec.

The action potential that triggers the heartbeat is generated within the heart itself. Motor nerves (of the autonomic nervous system) do run to the heart, but their effect is simply to modulate — increase or decrease — the intrinsic rate and the strength of the heartbeat. Even if the nerves are destroyed (as they are in a transplanted heart), the heart continues to beat.

The action potential that drives contraction of the heart passes from fiber to fiber through gap junctions. Significance: All the fibers contract in a synchronous wave that sweeps from the atria down through the ventricles and pumps blood out of the heart. Anything that interferes with this synchronous wave (such as damage to part of the heart muscle from a heart attack) may cause the fibers of the heart to beat at random — called fibrillation. Fibrillation is the ultimate cause of most deaths and its reversal is the function of defibrillators that are part of the equipment in ambulances, hospital emergency rooms, and recently on air lines.

The refractory period in heart muscle is longer than the period it takes for the muscle to contract (systole) and relax (diastole). Thus Tetanus is not possible. Cardiac muscle has a much richer supply of Mitochondria than skeletal muscle. This reflects its greater dependence on cellular respiration for ATP. Cardiac muscle has little Glycogen and gets little benefit from Glycolysis when the supply of Oxygen is limited. However anything that interrupts the flow of Oxygenated Blood to the heart leads quickly to damage, even death of the affected part. This is what happens in heart attacks. (89)

Smooth muscle (like cardiac muscle) does not depend on motor neurons to be stimulated. However, motor neurons (of the autonomic system) reach smooth muscle and can stimulate it, or relax it, depending on the neurotransmitters they release, such as Nitric Oxide (NO) or Noradrenaline. (48)

Before the latent period is over, the enzyme acetylcholinesterase breaks down the ACh in the neuromuscular junction (at a speed of 25,000 molecules per second) the sodium channels close, and the field is cleared for the arrival of another nerve impulse.

The resting potential of the fiber is restored by an outflow of potassium ions. The brief (1–2msec) period needed to restore the resting potential is called the refractory period.

Tetanus and Clonus:

The process of contracting takes some 50msec; relaxation of the fiber takes another 50–100msec. Because the refractory period is so much shorter than the time needed for contraction and relaxation, the fiber can be maintained in the contracted state so long as it is stimulated frequently enough (e.g. 50 stimuli per second). Such sustained contraction is called Tetanus.

Clonus and Tetanus are possible because the refractory period is much briefer than the time needed to complete a cycle of contraction and relaxation. Note that the amount of contraction is greater in Clonus and Tetanus than in a single twitch.

In the figure,
· When shocks are given at 1/sec, the muscle responds with a single twitch.

· At 5/sec and 10/sec, the individual twitches begin to fuse together, a phenomenon called Clonus.

· At 50 shocks per second, the (except for the cardiac) muscle goes into the smooth, sustained contraction of Tetanus. (48)

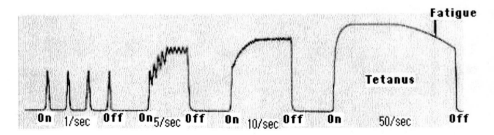

Image courtesy of J. Kimball

Ventricular Tachycardia and Ventricular Fibrillation:

A fast heart rate (over 100 beats per minute) on an Electrocardiogram is called a Monomorphic Ventricular Tachycardia. It means that all the beats match each other. When this is encountered by physicians, it is usually life threatening situation, calling for rapid diagnosis and treatment. (51)

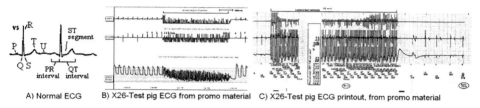

A) Normal ECG   B) X26-Test pig ECG from promo material   C) X26-Test pig ECG printout, from promo material

A better resolution graph was requested, but Taser International did not reply.

In graph **B** the lower ECG readout shows the blood pressure dropping during the 5 second firing sequence. The reason is the heart is pumping faster than the blood flow can supply. The Ventricles are not being allowed to fill fast enough before the next contraction occurs. The end result is a slowing or reduction of the blood flow.

Graph **C** shows the repeating sawtooth pattern of the ECG during another 5 second sequence. More on this will be explained in the Defibrillators section.

Cardiac pulse rhythm:

The rhythm of the heart beat is determined be the Sinoatrial (SA) Node. (8,52,85) Located within the heart, it receives input form the Central Nervous System (59) to increase or decrease the beats per minute of the heart. If the heart receives no input, then there is a backup system. In sequence order; S.A. Node, Atrial Foci, Junctional Foci, Ventricular Foci. The system works that the preceding Node or Foci must transmit a signal, and if it is not received within a certain time frame, then the next available Foci will trigger an action potential.

In the case of CEW's, they block the Central Nervous System by over stimulation, the regular communication signals get drown out, so the backup Foci of the heart are not activated, or are themselves drown out. The author can not find information that this possibility has been researched?

NMI Neuromuscular incapacitation:

The following three sections are taken from Taser International free literature.

The human nervous system communicates with simple electrical impulses. The command center (brain and spinal cord) processes information and makes decisions. The peripheral nervous system includes the sensory and motor nerves. The sensory nerves carry information from the body to the brain (temperature, touch, etc.). The motor nerves carry commands from the brain to the muscles to control movement.

TASER technology uses similar electrical impulses to cause stimulation of the sensory and motor nerves. Neuromuscular Incapacitation (NMI) occurs when a device is able to cause involuntary stimulation of both the sensory nerves and the motor nerves. It is not dependent on pain and is effective on subjects with a high level of pain tolerance.

Taser operation:

The Shaped Pulse is comprised of two pulse phases. The first phase, called the "Arc Phase" is optimized to generate a very high voltage to penetrate clothing, skin or other barriers. The "Arc Phase" is a very high voltage short duration pulse that can arc through up to 2 cumulative inches of clothing or barriers, or one inch per probe.

Once the arc is created, the air in the arc is ionized and becomes a low impedance electrical conductor that conducts the second pulse phase into the body. The second phase of the Shaped Pulse is the stimulation phase, or "Stim Phase."

The Stim Phase does not have to arc across a barrier, since this was accomplished by the Arc Phase. The Stim phase only has to flow across the highly conductive arc from the Arc Phase. Hence, the Stim Phase is optimized to provide maximum incapacitation for a human target while operating at super-efficient power levels.

Nerve muscles and stimulation:

The Neuromuscular system is quite lengthy and won't be shown in detail. The quick explanation is:

The CEW's override the Central Nervous System by a massive electrical charge. The analogy used was having two people talk on a phone, (the neuromuscular system) and having a third party interrupt and drown out the conversation (by the electrical charge). The analogy also states the network is not damaged by this process. A more detailed explanation of this process is on the Taser International web site under NMI (Neuro Muscular Incapacitation) Scientific Principals.

The human body runs on voltages of +100mv to -100mv (millivolts). CEW's deliver +1200-3000 (45) positive volts, *(nothing documented on negative volts but the waveforms show it - author)* into the subject. The only organ designed to resist these high voltages is the skin. The probes penetrate the skin, bypassing the only means of protection. The after effects of discharge is nerve tingling, the repeated firing of nerves. The author is unaware if this has been researched, however this can be triggering the heart to contract out of sequence, depending which nerve is being stimulated. (142) The cardiac muscle has backup protection to keep the blood pumping, but it only operates when there is no signal received.

Troponin:

Cardiac Troponin is a marker of all heart muscle damage. If a unconditioned cardiac muscle is over driven, Troponin would be the result. It is produced by several reasons, the most appropriate to this report is very heavy exercise or a marathon, defibrillation, and cardiac arrest.

Defibrillators:

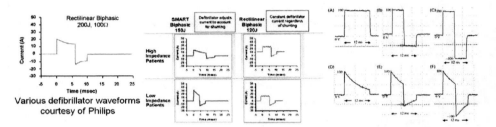

Various defibrillator waveforms courtesy of Philips

Biphasic Defibrillators (94) a device which delivers both positive and negative voltage and current, made by Philips. Philips literature states internal human resistance is between 25 and 180 ohms. Old model Defibrillators (that are no longer made) were just Monophasic, only delivered a positive pulse. The defibrillation waveform shares the same characteristics as the CEW waveforms, both positive and negative pulses, and less than a second timing phases. Monophasic is just a pulsed DC waveform. (34,35)

The exact shape of the waveform is not that important (unless purposely trying to defibrillate a patient), however any voltage rise above the action potential can trigger a response. (82,101) As previously stated in Tetanus and Clonus, any repeated electrical pulse beyond the action potential, faster than 5 to 10 pulses per second will be seen by a muscle as a clonus signal. (48)

The heart has a refractory phase where no matter the electrical charge, the muscle won't contract. After that the cardiac muscle is free to respond to the next pulse, while the rest of the body is in Clonus.

The heart not only is powered by a positive current, but a negative one as well. It became known as an Inward Current, or Funny Current. (97) The result was heart Defibrillators could now reduce the required output power and by delivering a positive wave followed by a negative wave (Biphasic) the same results could be duplicated with much less power. The power output is still higher than the CEW's can deliver, but there is also a wide margin involved because Defibrillators must work every time, on all body weights and compositions. The power delivered also has to fight through the skin and all layers of the body from known locations. (41,53)

An article titled "Disrupting the Hearts Tornado in Arrhythmia", Dr. Igor Efimov Ph.D. and his colleagues (19,32) say that state of the art Implantable Defibrillators use between 3-10 Joules to stop Arrhythmias. His research shows that if certain areas of the heart are targeted, successful defibrillation can be accomplished with only half-a-Joule of energy. The serious question can now asked, is if half-a-Joule, can stop an Arrhythmia, will more than that start one! (3) Compare with the waveforms on page 15, 49 & 59.

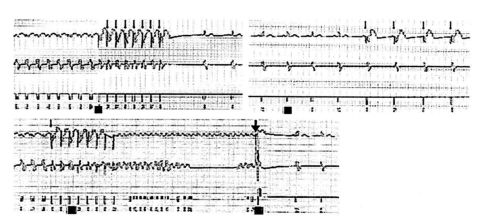

Waveforms of a pacemaker controlling the heart. Images courtesy of AAFP.

Note; In cases where patients have had pacemakers installed, the devices are tested by using a small voltage, which is less than a stun gun discharge, to trigger an arrhythmia condition. The pacemaker, to verify its working correctly will detect the condition and stop it.

Power output of the CEW's:

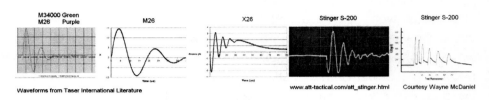

Waveforms from Taser International Literature · www.att-tactical.com/att_stinger.html · Courtesy Wayne McDaniel

This is the power output waveforms of the devices. The devices that have both a positive and negative output swing, are Biphasic, the last one is a pulsed DC, Monophasic waveform. The data sheets don't list what the negative swings are. (4,10,14,47) DC current would leave higher Acidity in the Blood Plasma. (70)

Taser International advertising quotes the M-26™ (13,14,15,16,17) as a 50,000 volt device, but only delivers 1200 volts into a person. (43,45) In the report for Taser International "The Advanced Taser, A Medical Review" by Anthony Bleetman (2003) he states the power output to be between 40,000 to 100,000 volts, resulting in a maximum of 3000 volts delivered into the target. (109) Note: Reference 45 says 5000 volts delivered?

In a follow on report on the "Taser X-26 Safety analysis". This report states that frequencies of 15-100hz and at power levels of of 500 milliamps are safe for use, and (usually) no organic damage is expected. This is for durations of 10 milliseconds or less only. Anything above that will cause burning.

The information on page 7 states in effect that that the frequency of the X26™ output is 100khz, but there is no experimental data or reported incidents using that frequency. The estimations of safety were extrapolated from graphing the 15-100hz and estimating their outputs at higher frequencies, as no data above 1000hz frequencies exist. The extrapolation gives a safe current to the body as 140 amps? The X26™ only delivers 50 Amps as a worst case scenario. The X26C™ is capable of 30 second firing times.

TASER® X26™ specifications: 26 Watts, 6000 volts peak, 1500 volts average, 18 pulses per second, 2.5ma average. (39,12,113)

Professor Shmuel (Sam) Ben-Yaakov in 2006 did a report for the Israel National Police. (140) "Electrical Evaluation of the M-26 Stun Weapon, Final Report" He reverse engineered the testing procedure and found faults. He clearly states "the manufacturers specifications are confusing, ambiguous and from a theoretical point of view, incorrect."

For example, manufacturers of electrical devices usually list on their data sheets minimum, typical, and maximum values. If a single number is given, it is assumed to be a maximum figure, such as 50,000 volts. As he shows, for an arc to cross a 35mm gap on the front of the device, it must be putting out a minimum of 70,000 volts. Using testing documents provided by Taser International *(not available to the public)* the 4 devices tested top current is between 6ma and 1.5ma at a 2000 volt input, and 100 ohm load. The data sheet just says "3.6 milliamp average", and their advertisements just states 3.6 milliamps. You can only convert Joules per second into a Watt, or a Watt per second into a Joule. Example; 1Watt per second = 1Joule. A Watt is Voltage multiplied by Amperage. 2000Volts x 6milliamp (.006) = 12Watts. 12 Watts multiply 5 (5 second activated trigger time) = 60 Joules. Current flow for 5 seconds is (1.5ma x 5 = 7.5ma) (3.6ma x 5 = 18ma) (6ma x 5 = 30ma)

The 1.76 Joules (as quoted in the advertising and data sheets) is at the main capacitor, not the final output. The main capacitor is also a confusing description. As another example, power regulators and amplifiers have a main capacitor and it's at the input side of the circuit, where the battery connects. It's intent is to store enough voltage to prevent the circuit voltage from dropping below a certain level while maintaining the current. Batteries are subject to dropouts and power fluctuations during rapid changes, the capacitor acts as a buffer so the battery supplies the capacitor and the capacitor supplies the circuit. For a 5 second pulse it should be 1.76 x 5 = 8.8Joules.

An advertisement for the X26™ on their website states, In previous generation pulse energy weapons such as the M26™, the pulse rate of the weapon would vary greatly depending on battery conditions. Particularly, in colder weather, the pulse rate could slow dramatically as battery performance decreased.

TASER® M26™ specifications 50,000-100,000 volts, 1.5-6ma current, 19 pulses per second, 1200-3000 volts delivered.

The Stinger S-200™ device specifications state 63,000 volts, 1300 volts delivered into the subject, at 5 milliamps, 17 pulses per second. Energy at main capacitor .60Joule. *The author can't find and independent testing on the device.* (46,49,56,95)
(Note: Instead of Joules, Taser International now uses micro coulombs as their unit of power measurement for their X3™ and newer models)

Electrostatic discharge:

Precipitation Static is the ability of the atmosphere or an object to hold an electrical charge, both positive and negative at rest. Electric motors carry electricity while in motion. Under the right circumstances the the static electricity can be forced to jump causing a spark or discharge.

The University of Florida and the US Air Force have a program in place to study lightning. This program is called "The International Centre for Lightning Research and Testing." and hosted at Camp Blanding. It consists of a small rocket that is launched into a storm cloud **(a)**. The rocket has a metal nose cone, that has a copper wire to it, and the other end is attached to the ground. The wire is insulated by a Kevlar sheath, so that only the nose of the rocket will conduct electricity.

Images courtesy of Chris Kridler
www.skydiary.com (108)

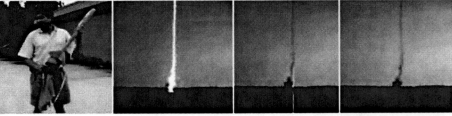

The Aviation industry teaches that any object traveling through an atmosphere will build up a positive electrostatic charge, that will continue to increase until discharged or released through static dissipation wicks. *2nd pic from right, is the rebound lightning charge.*

This charge can jump at levels less than 30 volts. (29,86) When launched, the rocket produces a positive electrostatic charge around itself, as it travels upward. This positive charge is capable of attracting negative electrostatic charges in the atmosphere, and if sufficient charges are available, they will jump, causing lightning and thunder. Lightning is the result of over whelming movement of electricity in the atmosphere. Thunder is the result of the charges jumping and forcing the air molecules around the charges to move aside at supersonic speeds. In electrical terms, two equal electrons (positive and positive, negative and negative) repel and opposites (positive and negative) electrons attract. The negative charges in the atmosphere are attracted by the positive charges built up around the rocket, travel down the wire until reaching ground, and most times rebounding back up.

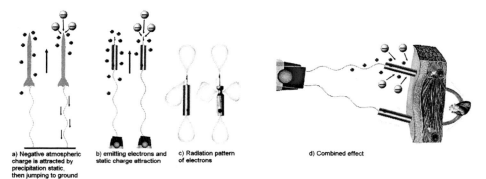

a) Negative atmospheric charge is attracted by precipitation static, then jumping to ground
b) emitting electrons and static charge attraction
c) Radiation pattern of electrons
d) Combined effect

On a smaller scale, air cleaner devices that are referred to as Ion Generators are just high voltage generators, usually in the 7000-13,000 volt range. They work on the principal that dust and other pollutants in the atmosphere are attracted to the static charges produced, and then get stuck on the filters of the machines. The devices are usually negatively powered, because dust has a positive charge from moving around in the atmosphere. The static charge while usually not enough to build up very much on dust to cause a visual display of lightning, you can hear the charges jumping as static sparks. The method that the generators use to disperse the high voltage is to funnel them through needles inside the air stream. As the electrons travel down the needles and reach the sharp end, this causes the electrons to spray outwards, similar to pinching the end of a garden hose while spraying water. This effect can be increased as the flow of air around the needle increases.

The reason is that electricity is attracted to the atoms **(a)** or molecules that make up the atmosphere, both positive and negative. Negative atoms have more electrons than protons. Positive atoms have less electrons than protons. Atoms can be either positive or negative depending on the amount of electron charges it gains or looses. An Ion is another name for a charged atom or molecule.

As the dart flies through the atmosphere, and a negatively charged ion travels past the head of the needle, it robes an electron from the charge building up on the probe, turning itself into a positive ion. The further the probe flies, the more positive charged electrons are laid down in a path, essentially seeding the atmosphere in the path of travel, just as the lightning rocket.

The CEW **(b)** devices that operate by launching probes at high velocities operate in a similar fashion. The unit builds up a high voltage charge when the probes are launched. As the probes travel through the air, the needle projectiles on the end would be emitting electrons as it travels **(c)**. The wires are shielded to prevent arcing with the grounding wire during flight, however the charges emitted just remain on or around the positive wire as they have no where to go. When the probes impact the target **(d)**, a closed circuit can now be made, meaning the electricity now has somewhere to go, and it also attracts the static charges that it laid down in its path of travel, as well as any attracting charges in the surrounding atmosphere. The electricity may just simply jump across the outside skin of the target to the grounded probe, but that as this hypothesis leads may also be through the blood stream.

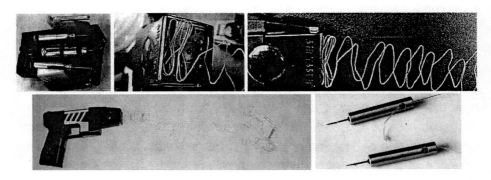

Upper images courtesy of R.T. Wyant. (12) Lower images courtesy of Sam Ben-Yaakov. (140)

Looking at these pictures, just ask yourself this question. Since the cartridges are not hermetically sealed, what is the sterility conditions of the probe needles? This basic question is overlooked, neglected, ignored or not even considered to be an issue. The majority of drug abusers in the world know better than to stick an unsterilized needle into themselves, and yet here is a device carried by the police forces (and civilians in some countries) that throw basic medical knowledge out the window. These devices are holstered down between the officers crotch and armpit, where it gets fluffed by who knows how many doughnut lunches and lavatory visits. There are no requirements for cleaning or sterilization procedures for these devices.

As the pictures show, there does not appear to be any insulation around the weighted probe (from either manufacturer), and arcing around probe has been observed, and is how the device operates. What the pictures don't show is that (X26™, M26™) where the cartridge plugs into the gun, there is a gap between those electrodes, so arcing must occur there first. Secondly arcing must occur at the probe where the wire connects. It is not permanently attached, it is only balled up so it doesn't pull out. The wire is insulated, however the arc must burn through the insulation before going to the probe. This intermittent connection is responsible for reported spikes in the power supply. This will not show on an ordinary oscilloscope, one needs a scope with a response time faster than 150 MHz to display this.

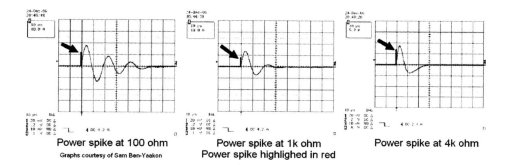

Power spike at 100 ohm   Power spike at 1k ohm   Power spike at 4k ohm
Graphs courtesy of Sam Ben-Yaakon   Power spike highlighed in red

For example; when turning on an incandescent light bulb, there is first a surge of current, and this surge is the reason that light bulbs almost always burn out when first turned on. Arcing would also cause a surging in an inductor, when first powered up, and when the arcing happens, that is when the surge happens. The probes were designed to first burn away the insulation around the wire, then ionization (96) of the air around the probe will reduce the power required to keep an arc sustained, and increase the flow of electricity through the probe. As the graphs show, the lower the resistance, the higher the power flow. These graphs were used to describe what was believed to be an indication problem in the testing oscilloscope, however they do represent the arcing that would occur in real use for the spark gap design. Graphs shown without the spike are documented as no-gap, meaning the cartridge was not attached. This is not representative of how it actually operates in the field.

Arcing from the probe tail to the subject has been observed in at least one video on the internet. Electricity will always follow the path of least resistance. Electricity running through a wire will not arc onto itself unless there is a path of least resistance, such as ionization around the darts. Otherwise stated as there is such a high volume of electrons surrounding the darts, that the path of least resistance through the air is less than the path through 1 inch of metal. The probe is made of two different metals, the body, and the needle, so this difference is causing a minor resistance. Even if the needles are penetrated into an object, this area where electricity jumps is also the attraction point for static electricity. Static electricity will jump if in contact with a metal object or an object of lower resistance, such as arcing.

Inductance of the wire:

An electrical current through a wire produces a magnetic field. This is also called a magnetic flux, and is the measurement of quantity of magnetism. The ratio of magnetic flux to current is called inductance. Inductance does not effect the steady flow of current, but resists the change when there is a change in current. This property allows wires shaped in a coil to temporarily store energy. (38,87)

When current is rapidly pulsed through a coiled wire, very high voltages can be produced, and this is how the CEW devices build up their 50,000 plus volt charge. Variable voltages can also effect the output. When the positive circuit pulse goes to ground or negative potential, static voltages at the probe end or wire attachment points are now able to enter the circuit and effect the inductor wires charging, keeping it more charged, and delivering higher current and voltage pulses when the primary charging circuit fires. Alternatively any negative static charges can be attracted to a positive charge.

Static electricity over a carpeted surface:

Many people are familiar with getting a static shock from walking over a carpet surface, or pulling clothes out of a dryer. Friction of materials generating a shock is one explanation, but this is also called the Triboelectric Effect. (69) It states that certain materials can become electrically charged after they come into contact with another different material, and are then separated. Human skin is one of the most positively charged items on the scale, and carpeting is a combination of positive and negative materials. This positive and negative difference means that electricity is able to move or be attracted between both, either by friction or held by adhesion.

For example: When you walk on a carpet with your bare feet, you (for the most part, based on body composition, weight, age) don't build up a static charge, it's able to dissipate through the carpet. When you wear shoes, you add an insulating factor, the friction of movement continues to build up the charge, but now the charge can't dissipate, so it grows around the body, and the body behaves like a capacitor storing the charge. When you come into contact with a grounded or conductive object, you get a shock from the electricity jumping, which is written to be in the thousands of volt range. Shocks of 2000 to 3000 volts or are not usually felt in most circumstances. Walking on a carpet can create several thousand up to the tens of thousands of volts. A search of the internet has even found voltages up in the 50,000 volts range for generating static charges on carpets. Semiconductor manufacturers have published many articles on Electrostatic problems and protection.

After being shot with the CEW, Mr. Dziekanski was retreating across the carpet, and the officer was following with him. Both would be generating static voltages during this time. It can then be stated that as previously described, the generated static voltages were attracted around the probes, and subsequently could have discharged through his body. The sound of the nitrogen capsule discharging and the arc phase was noticeably louder than other videos on the internet, possibly from the static. It was also louder than his yell from behind a quarter inch of glass. During the second shot, when he was lying on the ground, there was more surface contact, developing much greater voltages.

It should be noted that of the people being shocked by a CEW that launches probes during testing on any internet video, most do not scream out uncontrollably, or are able to move. Most go into a frozen state, are able to contain their outburst, or put out a vocal sound that you can tell is controllable. This was not the case here. This sequence of Mr. Dziekanski covers about 11 seconds. *(from the Pritchard video)*

Samples of the carpet from the YVR airport were obtained, showing the construction. The carpet is constructed of a heavy nylon weave, followed by two differing layers of fiberglass, and a dense rubber bottom. The adhesive material is unknown. This may be a durable construction method for high volume carpet wear, but the thick rubber also prevents any electrical charge from dissipating through it to the ground. Any static charge on the carpet, will remain there, and be generated. For static electricity to be generated the difference of Triboelectric materials are from bottom to top:

Rubber (negative), fiberglass (positive) adhesives (unknown, but usually negative), nylon (positive) – The rubber sole of a shoe (negative) and the human body (positive). What you have is the perfect combination for electrons to travel freely: - + - + - +

Images of the YVR airport carpeting; author.

Aircraft carpeting is a combination of usually a wool blend or solid nylon, both with a light rubber backing. Having too much of a wool content has caused frequent static build up and shocks to be encountered. Past manufacturing methods had fiberglass glued onto the carpet backsides. This resulted in an excellent wearing carpet that did not shrink or move due to aircraft flexing and temperature change. Although it was a little heavy and expensive, the current trend of airlines is to go with a cheaply manufactured carpet product with a light rubber backing. Instead of washing it, it is replaced every few months.

Influencing the output:

The probe wires are folded in a slot for storage in the cartridge, and unravel for deployment, the wires have been stored since manufacture and remain in that shape for many months / years until used. The wire even though is now deployed, would still behave as if it still was folded or coiled, and become a secondary inductor to the primary charging inductor. Even a straight wire has a small inductance to it, however the variability in the shape of the wire, the distance traveled and environment variables can change the output pulse. If the wire falls across a metal object or a power cable, it can induce a different output in the inductor than the primary charge circuit is designed to produce. For example, if the wire is spooled on a .5" form, multiply that by 35' 6" (the total length of the wires) and you get 852 coils. Wire size is 36AWG or .005 mil, this gives a possible of 1011uH (micro henry's) of inductance. (44) The RCMP in Canada as already covered provided no samples from the garbage bin to verify this.

The output power pulse can also be an attraction for the static electricity. The power swing goes from positive to zero to negative, so while at the zero to negative cycle the exposed electrical probe arc would allow positive charges to be conducting and charging the circuit. This becomes possible because the insulation around the wire is now burnt away, and more of the wire connection can now be in contact with the probe. The ionization would expand outwards and attract any opposing charges that are building up around the probe. The end result is that with additional static voltages charging the circuit, the gun would be pulsing out higher charges intermittently.

Condensation buildup within the unit itself, is another possibility not addressed in published testing of the CEW's. Condensation is a result of different materials, metal and plastic not being able to respond to changes in the surrounding atmosphere at the same time. Any condensation around the metal contacts within the plastic housing would reduce the arcing distance, and reduce the power required to jump the distance between contacts, leaving more powered delivered into target. Exposure to rain is another method the device can get wet inside.

The distance between the cartridge contacts is unknown. The stun gun goes from a temperature controlled office, outside, into the police car, sits in the holster next to the officers humid areas, then back outside. Sits holstered while in the rain, waiting to be used. This is the most likely situation every device will face. There does not appear to be any moisture protection or absorbing packets installed in the devices to prevent this from happening. All calculations for an arc jumping a gap is based on the air being dry and contamination free.

Combine at least the actions of, variable high voltages at both ends of inductor wire and the body acting as a capacitor, you would be left with the 35ft inductor wire outputting more Wattage than designed. This diagram is a representation of that, and the equivalent schematic of the circuit on the right.

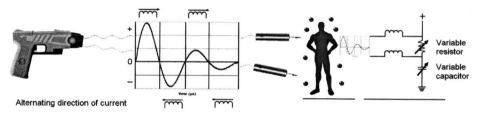

As previously stated the human body acts as a capacitor to hold a static charge, the rubber shoe bottoms hinders the easy flow of electrons, and movement through the air, plus friction on the carpet causes the voltages to increase substantially across the body. Static electricity will conduct to the nearest metallic object, or object of lower resistance, so therefor the probes would be conducting a current and voltage from that. It is conceivable that as each of the four officers came into contact with Mr. Dziekanski, they each added their static charge around them to the circuit. If the average voltage is 20,000 volts per person, then it would be 20,000 multiply by 5, as Mr. Dziekanski would also be part of the circuit. That power would be connecting down the probes and into him, overriding his nervous system, far longer than just the stun gun application alone, but directly contributed to by the device.

Even if the voltage was at the minimum jumping capability of 30 volts, this is still 300 times more voltage than the Central Nervous Systems + -100mv usual load.

Images from Pritchard video

In the picture on the opposite pages, this is the area where Mr. Dziekanski died. An electrostatic meter was constructed to test for both positive and negative static charges. The electrostatic meter shows that both positive and negative charges are found in this area, around the computer and electronic displays. His throwing of a computer could have resulted, in a rare instance of the capacitors discharging their voltages into the surrounding environment, increasing the local areas potential charge, acting similarly to a Van de Graaf generator.

In this prototype, the needle only going to midpoint was the maximum deflection that the meter was capable of. Further testing revealed that walking or sliding one's foot on the carpet was enough to maximize the meter's reading while holding it. This secure area was an average of 20 to 30 degrees warmer than the unsecured area and was often kept that way according the the security person stationed at the security door. The air in this area has poor circulation, and is dry from the air conditioning, perfect conditions for static electricity to exist.

None of the computer equipment in this area is switched off, it just remains on in power saver mode. The unsecured area to on the other side of the glass has exposure to the rest of the building, outside and roadway traffic, and much less of a chance for static electricity to build up there. The doorway and quarter inch glass provided an excellent insulation barrier between these two areas.

This is where the police stood as they faced Mr. Dziekanski.

Bottom Picture, is looking out to where Paul Pritchard was taking his video on the other side of the glass. *Images from author*

34 · CONDUCTIVE ELECTRONIC WEAPONS AND THEIR FAULTS

Behind the counter with a new computer and chair, facing the security doors.

Above and previous page; The meter is capable of detecting both positive and negative voltages. The upper meter reads positive and lower reads negative. This secured area was charged enough to give a fast full scale deflection. Appendix A has the construction details of the meter.

Author admits the exact needle position is hard to read due to limitations in printing resolution, he still has the meter should anybody want to test its capabilities. The MPF102 transistor specified in the circuit is getting hard to find now.

Excited Delirium:

Excited Delirium (9,99) is a confusing topic! The symptoms have been under study since 1849, yet not officially recognized as a separate medical condition. Death by Excited Delirium has no specific cause to any one branch of medical science. The medical system looks for signs or trauma to define a condition, but can't locate enough of a problem in any one field to say definitely what it is.

Excited Delirium is an example where there is no serious trauma to the body until it's broken and death happens. The body simply over exerts itself depending on the situation, and can't recover. It's the same situation as an engine that runs normally, then the governor breaks, and then it will destroy itself in short order without intervention.

Air Rage is another form of Excited Delirium. A passenger, flight attendant or pilot just suddenly goes berserk, sometimes suffers from hallucinations, has increased strength and is very hard to restrain. The restrained person on occasion has died.

Excited Delirium according to Morrison and Sadler (2001) is "A state of extreme mental and physiological excitement, characterized by extreme agitation, hyperthermia, epiphoria, hostility, exceptional strength and endurance without fatigue." (30)

It is the authors opinion that Excited Delirium is not the correct diagnosis when a person has been hit with a Conducted Electronic Weapon. It is the Self Preservation mechanism (131) that every animal has when exposed to pain and fear that determines their outcome.

There are similarities between the causes of Excited Delirium, Air Rage and Self Preservation when pain is involved. All involve some kind of Central Nervous System suppression, such as alcohol, medical drugs or illegal medication, and a reduction of Oxygen in the body.

Alcohol is a depressant on the Central Nervous System, also causes dehydration and leads to a higher acidity within the body. Drugs, both medical and illegal (pain killers or euphoric causing for example) specifically target the Central Nervous System. A persons health can be in a compromised state either from a sudden onset or from continual degradation. Conducted Electronic Weapons have been demonstrated to cause acidity, with a thirty minute rapid increase and thirty minute rapid decrease.

The bodies response to acidity is to remove it at the expense of other body functions, such as oxygen intake by the blood stream. A reduction of Oxygen intake causes deleterious effects to be experienced by the person. Some hypothesis go into more in depth body functions, however the author believes that is over analyzing. What controls the output of the body is the Cognitive abilities of the person.

What influences a persons judgment: Intoxicated and people under the influence of illegal drugs have a reduced sense of Cognition. Cognition is the ability of the body to interpret and recognize the physical stimuli that it is receiving. Sight, smell, touch, hearing, taste are the five senses. From that the person must decide what actions to take. If the Central Nervous System is suppressed, then the logical reasoning is impaired. Light impairment such as intoxication or drugs reduces judgment, inhibitions, reasoning, and minor motor reflexes. Heavy intoxication increases all of that and more.

The Fight or Flight response written about in many other reports, is an option an individual makes, but that also depends on what the stimuli is. Is the person clear headed in his thinking, or are they impaired? A threat, fear and pain are the primary influences. A perception of a threat, fear, and pain can trigger an increased responsive action to what is actually there. If the Central Nervous System is compromised, the body can't Fight, so it can only resort to Flight. The need to get away from a perceived or real threat is amplified when there is pain involved. (131)

Two recent examples of people under extreme duress. The Dziekanski video, and a passenger Air Rage from an Air Canada flight to Europe.

Images; Pritchard video, and the internet.

The Dziekanski video: As soon as the police go through the double doors, they essentially place him under arrest. He wants to know where they have been, he's been calling for them for a while, The officers again order him up against the counter, an officer pulls the stun gun, orders again, Mr. Dziekanski's body language essentially says, what the hell is this as he further tries to communicate, then the officer shoots. In extreme pain he tries to run away, the officer fires again, then they hold him down.

In Mr. Dziekanski's perception, the police are there to help, but when they indicate that he is under arrest, there is a sudden distrust, disappointment, and disbelief. The officer probably has his hand on the weapon, so the perception of fear and intimidation is set in. When he fires, the weapon suppresses the Central Nervous System, he can't think clearly, the need for self preservation takes over and the need to get away from the pain takes priority. The body would kick in adrenaline, he tries to run away but the officer fires again. From his bodies point of view, there is pain, confusion, suppression of movement from the CEW and the officers restraining him. The body is now in a state of terror, and believes that it is fighting for it's life. The body turns off the muscular feedback mechanism, which causes the super strength to be encountered. The static electricity entering his body from the probes would be acting as a sabotage mechanism on the bodies functions leaving it unable to control itself.

The recent airline passenger who went berserk was a situation that was handled differently. While he was also under restraint (he had an airline passenger restraining kit applied to his wrists) and two people on either side restraining him, no one was trying to attack him, rush him, nor was there anybody causing him pain.

One of the persons on the far side was even trying to calm him down. There was no need for the self preservation instinct to kick in, so no super human strength to fight off. He was yelling for water, and it probably would have reduced his anger if he was given water, and oxygen.

Positioning of the person has been cited as a contributing factor. The prone position is chosen by the police because it represents a position that takes an effort to get up from, and run away. The police will have a better chance of noticing and stopping someone trying to escape when using this position. This position is also used by the police and military snipers, however they are not completely flat and actually angle their bodies to reduce pressures. In situations involving an unconscious or intoxicated person, this position can cause Asphyxiation. The simple reason is the body weight is on the heart and lungs. When unconscious or intoxicated, the body does not have full control of the respiratory muscles.

The heart can operate on its backup system, but there is no such thing for the respiration. Respiration requires use of the Central Nervous System. Any kind of interference or suppression of that, such as a physical weight, medical sedation or anaesthesia, drugs such as cocaine, or poisons such as curare, can stop the respiratory muscles from functioning. Prolonged or continuous use of a Conducted Electrical Weapon would also keep the respiratory muscles in a state of Tetanus, causing a state of reduction of oxygen or loss of wind in the body. The X26C™ (for civilian use) is capable of repeated firings without intermission. Press the trigger three times and the device will continue to fire for 30 seconds. Police models don't have that feature?

First aiders are taught place an injured or unconscious person in the recovery position (134) because after a period of exertion, the body goes into a relaxation phase, muscles go limp and postural hypotension happens. (145) The chin is kept tilted upward so the Epiglottis does not close, (as it does in the prone position) (57) fluids are allowed to drain and gravity does not allow the tongue to fall back and cause blockage of the airway. Other functions such as control of fluids get compromised, and can enter the bronchial tree, causing aspiration

Pneumonia. (58) This can also happen during incorrect swallowing, and can happen to intoxicated people. Preoperative people are instructed not to eat before surgery because of this. The more acidic the foreign substance, the more sever the reaction.

Alcohol, Cocaine, Nicotine and Amphetamines:

At this time, there was no information given to the author, nor is he making any speculation to conclude that Mr. Dziekanski was on any medications at the time other than going through Nicotine withdrawal. The side effects of these drugs have been raised as a contributing factor to CEW deaths, they are included to support this hypothesis.

Alcohol (Ethyl Alcohol) (92) is a sedative that causes a suppression of the Central Nervous System. It affects the Central Nervous System by reducing the bodies neurotransmitter signal flow.

Cocaine (90) on the other hand has an excitatory effect. It causes significant damage to both the brain and Central Nervous System, increasing with usage. Cocaine increases the Dopamine (124) levels by blocking the transport mechanism, but the body continues the manufacture leading to excessive levels on the transmitters. It also increases the heart rate, blood pressure and breathing. It also acts as a local anaesthetic by blocking the sodium channels. Heavy cocaine users frequently have mood disturbances, paranoia, and auditory hallucinations. Chest pains, undernourishment, is also frequent and death is usually by Heart Arrhythmias and Heart Attacks, and Seizures. It acts, and was also used as an early anaesthesia (66,90) in medicine.

Nicotine (93), is also stimulant that increases the Dopamine levels, and the combination of Cocaine increases the risk of heart attacks. Cocaine users often chain smoke too, further increasing the dopamine levels. Withdrawal symptoms are; headaches, irritability and anxiety, difficulty concentrating and sleeping, decreased heart rate and blood pressure, and fatigue. (135)

The brain protects itself by either reducing the number of Nicotine Receptors or increasing them to disperse the toxins over a greater area. (136) *Much more is available at www.whyquit.com.*

Amphetamines (91) are a stimulant that increases the concentrations of Dopamine in the synaptic gaps. They also cause a release of Dopamine, Epinephrine (61), and Norepinephrine (123) into the blood stream and keep it working there for a long time.

Fight or Flight response:

First coined in 1915 by Walter Cannon (127), his theory states that animals react to threats with a general discharge of the Sympathetic Nervous System (126), priming the animal for fighting or fleeing. This response was later recognized as the first stage of a general adaptation syndrome that regulates stress responses among vertebrates and other organisms.

Adrenaline (Epinephrine) is under the control of the Central Nervous System. However the Sympathetic Nervous System can activate Adrenaline through the Splanchnic Nerves to the Adrenal Medulla (63). Two Catecholamines, Norepinephrine and Dopamine, act as neuromodulators (125) in the Central Nervous System and as hormones (129) in the blood circulation. The Catecholamine (62) Norepinephrine is a neuromodulator of the peripheral Sympathetic Nervous System but is also present in the blood (mostly through spillover from the Synapses of the Sympathetic System).

High Catecholamine levels in blood are associated with stress, which can be induced from psychological reactions or environmental stressors such as elevated sound levels, intense light, or low blood sugar levels.

Extremely high levels of Catecholamine (also known as Catecholamine Toxicity) can occur in Central Nervous System trauma due to stimulation and/or damage of nuclei in the brain stem, in particular those nuclei affecting the Sympathetic Nervous System. In emergency medicine, this occurrence is widely known as Catecholamine Dump.

Catecholamines have a half-life of approximately a few minutes when circulating in the blood. Monoamine Oxidase (MAO) is the main enzyme responsible for degradation of catecholamines. Monoamine Oxidases catalyse the oxidative deamination of Monoamines. Oxygen is used to remove an amine group from a molecule, resulting in the corresponding Aldehyde and Ammonia. If Oxygen is not available, the result takes longer to disperse.

The adrenal medulla is part of the adrenal gland. (128) It is located at the centre of the gland, being surrounded by the Adrenal Cortex. Composed mainly of hormone-producing chromaffin cells, the adrenal medulla is the principal site of the conversion of the amino acid Tyrosine into the Catecholamines Adrenaline (epinephrine), Noradrenaline (Norepinephrine), and Dopamine.

In response to stressors such as exercise or imminent danger, Medullary cells release Catecholamines into the blood in an 85:15 ratio of Adrenaline to Noradrenaline.

Notable effects of Adrenaline and Noradrenaline include increased heart rate and blood pressure, blood vessel constriction, bronchiole dilation, and increased metabolism, all of which are characteristic of the Fight-or-Flight response. Release of Catecholamines is stimulated by nerve impulses, and receptors for Catecholamines are widely distributed throughout the body. The Conducted Electronic Weapons nerve stimulation technology would also be capable of activating this stimulated response. This increased supply of blood pressure would as already stated reduce the bodies resistance to an electrical charge.

The Parasympathetic Nervous System uses only Acetylcholine (ACh) as its neurotransmitter. This transmitter is controllable by positive and negative voltages.

Anaesthesia, Sedation:

Anaesthesia by definition means a blocking of the sensations and pain. (60) It is standard practise that in any medical procedure on a patient that causes pain, anaesthesia be administered. Sedatives (64) are substances that depress the Central Nervous System, such as alcohol, and antidepressants. Anaesthesia and sedation causes relaxation of the respiratory muscles, and low blood pressures. A breathing apparatus and constant monitoring is required for any medical procedures that uses either to get the subject into unconsciousness.

Adverse effects of local anaesthesia are generally referred to as Local Anaesthetic Toxicity. Local anaesthetic drugs are toxic to the heart (where they cause Arrhythmia) and brain (where they may cause unconsciousness and seizures). Arrhythmias may be resistant to defibrillation and other standard treatments, and may lead to loss of heart function and death.

The first evidence of local anaesthetic toxicity involves the nervous system, including agitation, confusion, dizziness, blurred vision, tinnitus, a metallic taste in the mouth, and nausea that can quickly progress to seizures and cardiovascular collapse.

Toxicity can occur with any local anaesthetic as an individual reaction by that patient. Possible toxicity can be tested with preoperative procedures to avoid toxic reactions during surgery.

Inhaled general anaesthesias such as Nitrous Oxide requires additional help to induce unconsciousness, however any Central Nervous System suppressant supplies that help. Nitrous Oxide is naturally produced by the body to relax muscles, and is also a possible re-combination that can happen when the Electrolysis process splits the water molecules in the blood plasma. As stated on page 10-11, as the circulation flows through the body, and depending on where the probes struck, there can be areas of heavy concentrations of Nitrous Oxide suddenly spiking the weakest organ.

The most popular inhaled Anaesthesias are Isoflurane (116), Sevoflurane (117), and Desflurane. (118) Isoflurane is in decline in use on humans because of recent testing links to Alzheimer's disease, but remains strong Veterinary use. The patent for it has expired, so cost wise is economical to use.

Several of the researchers were queried for the types of anaesthesia they used to eliminate them as suspects for causing testing irregularities. While it is generally understood that all anaesthesia's result in sedation, how each one specifically operates is different, and may result in differences of testing results.

Dr. Walter's team used Ketamine (114) and Xylazine. (115)

Ketamine is a dissociative anaesthetic for use in human and veterinary medicine. Its hydrochloride salt is sold as Ketanest, Ketaset, and Ketalar. Pharmacologically, Ketamine is classified as an NMDA receptor antagonist, and at high, fully anaesthetic level doses, ketamine has also been found to bind to opioid mu receptors and sigma receptors, and like other drugs of this class such as Tiletamine and Phencyclidine (PCP) induces a state referred to as "dissociative anaesthesia."

Ketamine has a wide range of effects in humans, including analgesia, anaesthesia, hallucinations, elevated blood pressure, and bronchodilation. It is primarily used for the induction and maintenance of general anaesthesia, usually in combination with some sedative drug.

Ketamine, like Phencyclidine (130), is primarily a non-competitive antagonist of the NMDA receptor, which opens in response to binding of the neurotransmitter glutamate. This NMDA receptor mediates the analgesic (reduction of pain) effects of ketamine at low doses. Evidence for this is reinforced by the fact that Naloxone, an opioid antagonist, does not reverse the analgesia. Studies also seem to indicate that ketamine is "use dependent" meaning it only initiates its blocking action once a glutamate binds to the NMDA receptor.

At high, fully anaesthetic level doses, Ketamine has also been found to bind to opioid mu receptors and sigma receptors. Thus, loss of consciousness that occurs at high doses may be partially due to binding at the opioid mu and sigma receptors.

Xylazine is a drug that is used for sedation, anaesthesia, muscle relaxation, and analgesia in animals such as horses, cattle and other large non-human mammals. An analogue of Clonidine, it is an agonist at the α2 class of adrenergic receptor. As with other α2 agonists, adverse effects include bradycardia, conduction disturbances, and myocardial depression.

In veterinary anaesthesia, Xylazine is often used in combination with Ketamine. No formal information in humans available. Dr. Walter reported that inhaled Isoflurane was also used in a few animals and showed the same rapid cardiac results.

Dr. Jauchen used Ketamine and Propofol. (122)

Propofol is a short-acting intravenous anaesthetic agent used for the induction of general anaesthesia in adult patients and paediatric patients older than 3 years of age; maintenance of general anaesthesia in adult patients and paediatric patients older than 2 months of age; and sedation in medical contexts, such as intensive care unit (ICU) sedation for intubated, mechanically ventilated adults, and in procedures such as colonoscopy. It provides no analgesia.

Propofol is highly protein bound in vivo and is metabolized by conjugation in the liver. Its rate of clearance exceeds hepatic blood flow, suggesting an extrahepatic site of elimination as well. Its mechanism of action is uncertain, but it is postulated that its primary effect may be potentiation of the GABA-A receptor, possibly by slowing the channel closing time. Recent research has also suggested the endocannabinoid system may contribute significantly to Propofol's anaesthetic action and to its unique properties.

The elimination half-life of Propofol has been estimated to be between 2–24 hours.

However, its duration of clinical effect is much shorter because Propofol is rapidly distributed into peripheral tissues. When used for IV sedation Propofol typically wears off in minutes. Propofol is versatile; the drug can be given for short or prolonged sedation as well as for general anaesthesia. Its use is not associated with nausea as opposed to opioid medications.

These characteristics of rapid onset and recovery along with its amnesic effects have led to its widespread use for sedation and anaesthesia.

Aside from the hypotension (mainly through vasodilatation) and transient apnea following induction doses, one of Propofol's most frequent side effects is pain on injection, especially in smaller veins. This pain can be mitigated by pretreatment with lidocaine. Patients tend to show great variability in their response to Propofol, at times showing profound sedation with small doses.

Another recently described rare, but serious, side effect is Propofol infusion syndrome. This potentially lethal metabolic derangement has been reported in critically-ill patients after a prolonged infusion of high-dose Propofol in combination with catecholamines and / or corticosteroids

Abuse of Propofol as a recreational drug has been reported, usually among medical staff such as anaesthetists who have access to the drug. Despite a lack of analgesic properties, Propofol's sedative action presumably produces euphoric effects. The steep dose response curve of the drug makes such abuse very dangerous without proper monitoring, and several deaths have been recorded.

Dr. Webster and his team used Fluranes. Dr. McDaniel's used Telazol (119,120,121) and Isoflurane. (116)

The mechanism of action of Isoflurane;

Isoflurane reduces pain sensitivity (analgesia) and relaxes muscles.

The mechanism by which general anaesthetics produce the anaesthetic state is not clearly understood but likely involves interactions with multiple receptor sites to interfere with synaptic transmission. Isoflurane binds to GABA receptors, glutamate receptors and glycine receptors, and also inhibits conduction in activated potassium channels. Glycine inhibition helps to inhibit motor function, while bonding to glutamate receptors mimics the effects of NMDA. It activates calcium ATPase through an increase in membrane fluidity, and binds to the D subunit of ATP synthase and NADH dehydrogenase. In addition, a number of general anaesthetics attenuate gap junction communication, which could contribute to anaesthetic action.

Telazol (Tiletamine) is a dissociative anaesthetic and pharmacologically classified as an NMDA receptor antagonist. It is related chemically and pharmacologically to other anaesthetics in this family such as Ketamine and Phencyclidine. Tiletamine hydrochloride exists as odourless white crystals.

It is used in veterinary medicine in the compound product Telazol (Tiletamine / Zolazepam, 50mg/ml of each in 5ml vial) as an injectable anaesthetic. It is sometimes used in combination with Xylazine (Rompun) to tranquilize large mammals such as bears and horses. Telazol is the only commercially available Tiletamine product in the USA.

Contraindicated in patients of an ASA statues of III or greater and in animals with CNS signs, hyperthyroidism, cardiac disease, pancreatic or renal disease, pregnancy, glaucoma, or penetrating eye injuries.

Telazol is a proprietary combination of two drugs, a dissociative anaesthetic drug, Tiletamine, with the benzodiazepine anxiolytic drug, Zolazepam.

Tiletamine belongs to a class of drugs known as dissociative hypnotics, and is similar to Phencyclidine (PCP). It works by disrupting the Central Nervous System and induces a cataleptic state.

This drug does not provide muscle relaxation.

Zolazepam alone provides only subtle evidence of its presence, unless high doses are given. However, when combined with Tiletamine, a composite state of immobility, muscle relaxation, freedom from reflex movement, and analgesia prevails. This state provides conditions suitable for various diagnostic and therapeutic interventions, as well as for certain types of surgery.

Taser International had a brief statement on their web site before removing it *(found while searching the web, and recalling from memory)* that essentially stated that "It was able to prove that by blocking certain receptors of the Central Nervous System that its devices were not leading the cardiac muscle"

It would be reasonable to conclude that any substance such as anaesthesia, blocking the the bodies communication system would also block, (or theoretically increase the resistance of the body) the transmission output by the CEW's. It is also very likely that if the CEW under such circumstances would not be delivering it's intended output charge because of the theoretical higher resistance / blocking would not show the same results as a un-sedated person or animal.

Is is also reasonable to conclude that a person who is on illegal drugs, prescription medication, or other mechanism of suppression of the Central Nervous System that an application of electricity by a CEW would cause an increase in the already present Catecholamine levels. Procedures that cause fear, intimidation, and pain would activate the self preservation mechanism of the body and super human strength and the after effects that go with it.

The body can not always handle the rapid changes that goes with the self preservation response. The reason is that this response by the body is the same as any panic button. It is designed as a last resort to shut down all non essential systems including the bodies digestive, muscular feedback and immune mechanism. It is the cognitive perception by the person, even under the influence of medicinal or illegal medication is that if it is in danger, pain, or being restrained, means that it will activate the self defense mechanism to escape.

The body can not always recover from this and Heart Arrhythmias happen frequently while in the recovery mode, from the stress and strain of the event.

Long term effects of electrocution injuries:

Other than direct physical damage to the body by the CEW's the long term known effects of electrocution injuries are, muscle fibrosis, peripheral neuropathology, joint stiffness, reflex sympathetic dystrophy, cataracts, Central Nervous System disorders, subtle mental changes, and memory loss. (25,32,67,102,103,104)

Just as in any electronic machine, a voltage that goes above the rated carrying capability of the medium that it is traveling in will do damage. The body can repair most physical trauma to organs, it can not repair the Nervous System, once its damaged, that's it. There are many parts of the Nervous System that are just too specialized in their functions. (103)

The author can find no testing on what the voltage or current rating is of the Central Nervous System? It naturally handles **+ -100millivolts**, but there is no information on maximum safe voltages.

Note: In the animal slaughter industry, it has been found that electrical frequencies more than 800 hertz will cause damage to the meat in pigs and sheep (blood spattering or massive petechiae). See Appendix E: Comparison of Devices for operational frequencies of the stun guns. See also the update on page 190.

Chapter 2 Update April 12, 2008

Stinger System requested (after this report was sent out for review) that the three documents provided be reviewed for information.

1) White Paper Waveforms (written by Stinger Systems) essentially states why their device may be a better unit than the Taser International devices. For some reason the X26™ graph is marked as the X46, and the M26™ is marked the M46? The power graphs show the direct contact and gap methods for measuring high voltages. The testing procedure is explained in another report "Electrical Evaluation of the Taser M26 Stun Weapon Final Report, by Shmuel Ben-Yaakov".

2) In the document "Cardiac Safety of the Surface Application of the Stinger S200", Dr. McDaniel's applies 216 charges to 6 pigs. The animals were sedated and had an assisted breathing apparatus installed.

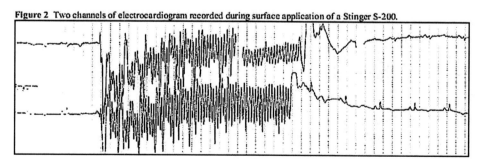

Figure 2 Two channels of electrocardiogram recorded during surface application of a Stinger S-200.

That report provides this graph as a typical example of an application, you can see that the S-200™ drives the electrocardiogram recording, very much like the other graphs on page 9 *(of that report)*. The ECG machines are very sensitive to electronic pulses and CEW's in most cases override their sensitivity just as they override the muscles of the test subject. The report states, "From this typical graph it was determined that no VF or VT had happened, using a 12 lead device".

3) The 2 page document on "Current Levels of Projectile Stun Guns Applied to Swine"

Device	Orientation	Peak Current (A)	RMS Current (mA)	Mean Current (mA)
Stinger S-200	S-S	1.96	43.1	2.6
Stinger S-200	SN-X	1.88	40.1	2.2
Stinger S-200	SN-Umb	2.12	40.5	2.2
Taser X-26	S-S	3.48	50.2	1.0
Taser X-26	SN-X	3.40	55.7	1.3
Taser X-26	SN-Umb	3.64	52.3	1.3
Taser M-26	S-S	14.6	147	0.5
Taser M-26	SN-X	15.3	137	0.5
Taser M-26	SN-Umb	15.3	145	0.6

S-S — Side to Side across heart
SN-X — Sternal notch to Xiphoid
SN-Umb — Sternal notch to Umbilicus

From these numbers, you can see the M26™ Mean Current is unusually less than the other two devices, considering the RMS or Peak current the device puts out.

Flat batteries can affect Taser records
The West Australian April 21, 2010

An internal police investigation has revealed that electronic records of the use of Tasers can be corrupted if the batteries in the devices are not properly monitored.

The investigation into the death of 49-year-old Mark Lewis Conway at Fremantle in August 2007 found that some records of activation stored on a Taser being carried by one of the arresting officers were inaccurate because its battery had gone flat, resulting in its time clock resetting to November 30, 1999.

Although the Taser was not used, the time and dates of some of its records of use were inaccurate.

The findings of the investigation stated that WA officers are advised that Tasers should not be used once their battery power, which is listed on their main display screen, falls below 20 per cent because of the potential for "brown outs" - the corruption of the activation records which are electronically stored in the devices. (1)

Chapter 3 Questions and Answers

Q: If the CEW's block the electrocardiogram from reading the heart, then how can medical professionals read the charts and determine that no fibrillation or tachycardia has taken place?

A: Electrocardiograms are simply voltage measuring devices, and if too much voltage is applied it will overdrive the measuring circuitry. From the author's non medical point of view, the graph on page 48 is non readable. At best, if the peaks of the graph represent the same peaks of the cardiac cycle, then the closer the peaks are to each other would suggest the heart is being driven faster during the application. The ECG was not designed to keep up with the CEW pulse cycle and is being thrown around by the discharge. As covered by reference 28, one doctor opened the chest of a pig to visually view the heart, and did see fibrillation, which the animal died shortly later.

Q: One researcher believes Lactic Acid is the leading culprit for the Acidosis?

A: Lactic Acid has been found in the testing and autopsy results, and is covered in this author's hypothesis, however Lactic Acid alone does not quite explain all the results seen, or the mechanism of action, or by itself produce Acidosis. (144) As stated in the BCOPCC final report; The Acidosis has a unique pattern of a 30 minute increase time and a 30 minute decrease.

Lactic Acid is a natural process of Rigor Mortis, and starts about 3-4 hours after death and lasts for about 24 hours. The time line for that mechanism does not match. There is also no explanation as to how the body gives up so much oxygen to get into this state. The author states in his hypothesis that, Electrolysis is the process that can immediately liberate oxygen to start this reaction.

Q: Is there more than one type of Acidosis?

A: Yes, and the author believes that more than one type of acidic condition comes into play. Medical science and other researchers are looking for a culprit, however if it came down to pointing out one single culprit for all the symptoms of Acidosis and unexplained deaths, medical science would have found that long ago. If one single factor can not explain the events as documented, then it falls to multiple factors being the cause, and that is the million dollar question. The author believes that Carbonic Acid forms first, and Lactic Acid is a secondary reaction to the body giving up Oxygen in the Red Blood cells to eliminate the Carbonic Acid.

Q: On page 10 you state that "A Oxygen / Hydrogen reaction should also cause some minor heating?" This has never been medically documented or simulated before?

A: Yes, this is not based on any medical experience, it is based on the author's previous experience with rocketry, and fuel cells. When you separate and rejoin Oxygen and Hydrogen molecules, you are using energy to do this, and the energy can only be released by creating some minor heating. The author does not believe it is very much, or even if it would be noticeable by the subject due to the pain caused by the CEW. In the search for answers it is clear that every clue is being utilized and as other researchers are doing, documenting as much as possible as it may someday be beneficial for future research.

Q: On page 16 under "Cardiac Pulse Rhythm," you state that "In the case of CEW's, they block the Central Nervous System by over stimulation, the regular communication signals get drown out, so the backup Foci of the heart are not activated, or are themselves drown out. The author can not find information that this possibility has been researched?"

A: The CEW's operation as has been shown can override the pacemaker ability of the heart and can lead it. The hearts backup mechanism, the Atrial Foci, Junctional Foci, and Ventricular Foci, will only take control of the heart when there is no signal received from the preceding S.A. Node or Foci's in that order.

1) The author can not find any information if research has been done to study if the backups of the heart are damaged by this leading process.

2a) Since the Electrocardiograms have a very difficult time detecting this signal during an application of a CEW, do the Foci (as a function) assume that the heart is still under normal control or is there a mechanism that comes into effect where the Foci try to take control of a runaway heart and slow it down?

2b) Can these devices damage or injure the Foci in that process, that when they are required to take over the heart in an emergency, that their injury could go undetected, and not work or work correctly?

Q: On page 17 you state; "The after effects of discharge is nerve tingling, the repeated firing of nerves. The author is unaware if this has been researched, however this can be triggering the heart to contract out of sequence, depending which nerve is being stimulated. The cardiac muscle has backup protection to keep the blood pumping, but it only operates when there is no signal received."

A: Nerve tingling is a side effect of a CEW application, caused by the movement of high voltage electricity across the nerves. The question; is it possible for tingling nerves in the body, or the heart specifically to somehow falsely activate a cardiac cycle, or interfere with the normal cardiac rhythm, causing a Fibrillation condition? *The author does not know if anybody is even researching that?*

Q: On page 22, you state that voltages can jump as low as or less than 30 volts, but air breaks down at 30kV/cm? Even Paschen's Law gives a minimum of 380 volts as the lowest possible level a static spark can jump?

A: This is, and is not a trick question. Everyone is familiar with having an electric shock from being on a carpet and touching a metal object, or pulling clothes out of a clothes dryer. Sometimes there is a gap between you and the object, and sometimes you have to seemingly touch the object first. As previously stated, 2000 volts is about the limit of being able to detect a static shock through skin. Below that and the voltages will just transfer through contact or adhesion, without the subject being aware.

In the semiconductor industry, some of those microchip devices in years gone by were extremely sensitive to any static electrical charge, such that they could be damaged when the voltage rises above 30 volts, and in some cases 12 volts. In the airline industry United Airlines wrote an avionics training manual (in 1974) and they listed 30 volts or less as the lower limit that avionics can be damaged by static electricity, and this where the author got that number from. That is what has been taught in the airline industry since then in the Jeppesen Aviation training manuals.

In the example used on page 31, 30 volts is just an example the author used to describe how each officer, by following the procedure of restraining, could have inadvertently delivered the static charge built up around each of them into Mr. Dziekanski. If the number was 30,000 static volts around each of them, then the charge delivered would be 300,000 times more load than the Central Nervous System's +-100mV normal load, and ten times more than the 1200-3000 volts normally delivered by the CEW's.

Q: On page 25 you state that "Secondly, arcing must occur at the probe where the wire connects. It is not permanently attached, it is only balled up so it doesn't pull out. The wire is insulated, however the arc must burn through the insulation before going to the probe." Doesn't that contradict your arguments on pages 22 - 24?

A: No it doesn't. In the pictures to the right, the one on the left is a commercially available negative ion generator. The center is an aircraft static wick for dissipating lightning and static voltages from any aircraft, and the right picture is the probes from a TASER® M26™.

As stated on page 23, "As the electrons travel down the needles and reach the sharp end, this causes the electrons to spray outwards, similar to pinching the end of a garden hose while spraying water. This effect can be increased as the flow of air around the needle increases, or the power at the needle increases. As you can see in the pictures, there is no difference if the electrons travel down a pin or a wire. Airflow past the head of either will drain off electrons.

In the left picture the ion generator needles are usually aligned in a perpendicular direction in the low speed airflow. In the center picture the static wick, the forward direction is to the upper right and the needle trails in the airflow. The yellow bumpers are for protection of the needle. The probe is not insulated where the wire was cut, and this allows the leakage of electrons. When the probes contact the object the leakage electrons allow the full charge to rush through and engage the arcing process. There must be a path for electrons to follow otherwise if the wire was completely insulated, the device couldn't discharge and would not be able to work.

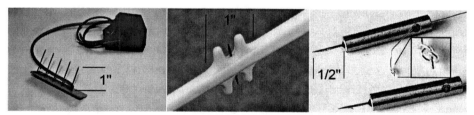

Images from: Author, Wikipedia and Sam Ben Yaakov

Q: Can the electricity emitted by the probe really leave electrons surrounding the wire?

A: Yes, this is the same principal of operation as the Van de Graaff generator. Electrons can travel and exist on a non conductive surface. Look up "Van_de_Graaff_generator" in Wikipedia for more information.

Q: Can you show how the inductance of the cartridge wire affects the output?

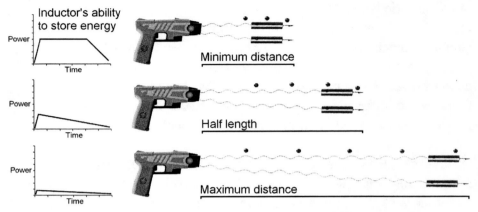

As explained on page 26, 27 & 30, the wire is folded into the cartridge, and carries the highest inductance at that point. After being fired, the inductance will be reducing because the wire is unfolding out of the cartridge. During this time, the CEW is keeping the primary charging circuit powered to it's fullest, and the folded wire is acting as extra storage capacity. As the inductance reduces, so does the ability of the wire to hold the extra charge. As the probes travel through the air, the extra charge would be emitted as electrons (as well as the devices pumped 50,000 volts), as previously explained on pages 24. How the different cartridges and lengths of the inductor wire would be a concern is the police issue cartridge (35' 6") is over twice the length and would have over twice the holding capacity of a standard cartridge (15' 6"), which is the type U.S. citizens have access to. *Update on page 66.*

Q: You list the recovery position or with legs elevated, as the position to place the person after being shot?

A: The recovery position provides the best hope for an individual whom is intoxicated, or under the influence of illegal drugs. If the individual is resisting arrest, is able to stand up right afterwords then obviously it's not required. When an individual is unable to move because one of the afore mentioned drugs or alcohol, or is unconscious, should be placed in the recovery position, if no medical personal are on site. This is in keeping with standard first aid responder practises.

As getting hit by a stun gun has been related to running a marathon on the heart. This is a similar condition to Runners Hypotension, when they collapse at the finish line of a race. The practice there is to keep moving or elevate the legs because the blood is pooling in the legs causing a sudden blood pressure drop. (145)

Q: Can you provide a graph to show how your hypothesis on acidosis works, after a stun gun discharge?

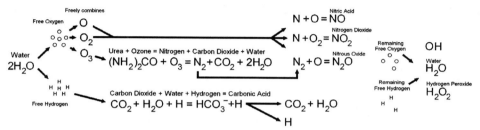

This shows the list of possible reactions in the human body once electrolysis of the water has started.

To meet the 30 minute increase in Acidosis, the newly liberated oxygen wants to recombines and the easiest combination is O_2, and then will also want to recombine again to O_3, Ozone. Ozone will only stay in that state for no more than 30 minutes.

During this time, the liberated Hydrogen will combine with the Carbon Dioxide, and mix with unaffected Water within the blood supply. This combination produces Carbonic Acid. Carbonic Acid will only remain in that state with the one Hydrogen molecule attached. This is the Acidosis increasing phase.

Anything that removes the Hydrogen from the Carbonic Acid will cause it to break down back into Carbon dioxide and Water. After the 30 minutes is over, the Ozone has reverted back to O and O_2. It should be stated that if there is no urea to mix with, then the Ozone will remain there for duration of the time span.

The body primarily removes Carbon Dioxide through the Red Blood Cells and exhalation of breath so this process continually reduces the available Carbon Dioxide. Without the Ozone molecule, Urea doesn't decompose allowing that component add to the Carbonic Acid cycle. The decomposed Urea, will primarily be robbing a Hydrogen element from another Carbonic Acid Molecule, but will also combine with a free Hydrogen molecule to form another Carbonic Acid molecule. At this stage the acidity would be in a reduction phase.

The liberated Hydrogen does not want to freely combine with anything especially itself, and this process will not reverse itself equally. The amount of remaining Oxygen should be greater than the remaining amount of hydrogen. As the body cycles the blood, it is possible for the remaining Oxygen and Hydrogen to form into combinations of OH, Oxygen / Hydrogen, Water and Hydrogen Peroxide, which is acidic.

This time limited chemical reaction has a wide variability of effectiveness due to the variability of concentrations of fluids and molecules. Effectiveness relies on the penetration of the CEW probes, the path of electricity, strength of electrical signal and affected fluids through the body.

After being hit by a stun gun, do you smell ammonia?

The chemical make-up of ammonia is nh3 (nitrogen and hydrogen). This means that there is one nitrogen atom bound to three hydrogen atoms. Ammonia can be a weak acid or a weak base, depending on what type of chemical it is suspended in. Ammonia has a strong, pungent odor that is easily recognizable in cleaning products, cat urine, and, for some people, sweat! (body builders smell it frequently.)

The key to ammonia in urine and sweat is the nitrogen. The body does this by stripping the nitrogen atom off of the molecule. The skeleton molecule that is left behind is then further converted into glucose and used as fuel. In order to get rid of the excess nitrogen, your body typically processes the nitrogen in your kidneys and forms urea, $CO(NH_2)2$ basically, a carbon dioxide molecule bound to nitrogen and hydrogen. Urea is then excreted in the urine. If your kidneys cannot handle the load of nitrogen, then the nitrogen will be excreted as ammonia in your sweat. (1) This is another side effect.

Chapter 4 Update August 17, 2008

The Thomas Braidwood inquiry has been divided into two parts, the first part covers the police procedure and the second part covers the events around Mr. Dziekanski's death.

While reviewing other presenters submitted information the author has found corroborating medical evidence to support his submitted evidence. Dr. Christian M. Sloane made a presentation from the University of California, San Diego. His report covered "Monitoring of human subjects pre and post exposure to the Taser." The date on this graph is December 2004. (3)

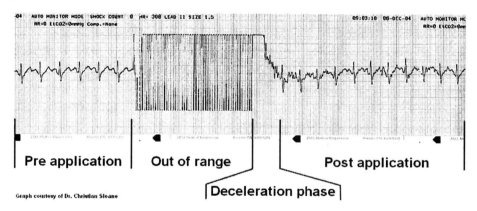

Graph courtesy of Dr. Christian Sloane

This graph shows the **Pre application**, the normal heart rate before application.

The **Out of range** which is the application of the stun gun voltage throwing the electrocardiogram meter against the upper and lower range scale. Electrocardiogram machines as previously stated in the author's report are designed to accurately graph the hearts + -100 millivolt range. The 1200 to 3000 volts applied by the gun is 12,000 to 30,000 times greater than the electrocardiogram's normal ability to accurately display the results. Electrocardiograms were designed to look for the bodies 100millivolt signal levels.

The **Deceleration phase** is the period after the high voltage application has stopped. The meter does not immediately return to normal because the body is still reacting to the after effects of the high voltage discharge. There is also the circuitry of the electrocardiogram possibly causing a minor delay as it tries to reacquire the correct voltage of the heart. This would be different for each machine.

The **Post application** is the electrocardiogram reacquiring the heart beat. As can be seen, the **Pre application** and **Post application** graph's don't match, therefore the cardiac rhythm has been changed by the use of the stun gun.

In Dr. Sloane's presentation he states that "A significant increase in the heart rate was found after a brief shock from a CED (Conducted Electronic Device), the Taser X-26". He then goes on to state that a few subjects have QT (specific cardiac cycles of the heart) changes, "The significance of which is unclear." Since he does not further explain the conditions or show a graph of QT changes, those known conditions are presented next. The image below is of a typical textbook sinus rhythm of the heart. *Image from Wikipedia.*

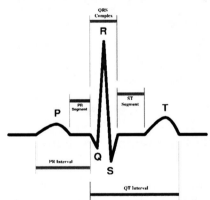

Short QT syndrome is a genetic disease of the electrical system of the heart. It consists of a constellation of signs and symptoms, consisting of a short QT interval interval on EKG (≤ 300ms) that doesn't significantly change with heart rate, tall and peaked T waves, and a structurally normal heart. Short QT syndrome appears to be inherited in an autosomal dominant pattern, and a few affected families have been identified.

The diagnosis of short QT syndrome consists of characteristic history and findings on EKG and electrophysiologic testing. There are currently no set guidelines for the diagnosis of short QT syndrome.

The characteristic findings of short QT syndrome on EKG are a short QT interval, typically ≤ 300ms, that doesn't significantly change with the heart rate. Tall, peaked T waves may also be noted. Individuals may also have an underlying atrial rhythm of atrial fibrillation.

The etiology of short QT syndrome is unclear at this time. A current hypothesis is that short QT syndrome is due to increased activity of outward potassium currents in phase 2 and 3 of the cardiac action potential. This would cause a shortening of the plateau phase of the action potential (phase 2), causing a shortening of the overall action potential, leading to an overall shortening of refractory periods and the QT interval.

Currently, the only effective treatment option for individuals with short QT syndrome is implantation of an implantable cardioverter-defibrillator (ICD). (5)

The long QT syndrome (LQTS) is a heart condition associated with prolongation of repolarization (recovery) following depolarization (excitation) of the cardiac ventricles. It is associated with syncope (fainting) and sudden death due to ventricular arrhythmias. Arrhythmias in individuals with LQTS are often associated with exercise or excitement. LQTS is associated with the rare, ventricular arrhythmia torsade de pointes, which can deteriorate into ventricular fibrillation and ultimately death.

Individuals with LQTS have a prolongation of the QT interval on the ECG. The Q wave on the ECG corresponds to ventricular depolarization while the T wave corresponds to ventricular repolarization. The QT interval is measured from the Q point to the end of the T wave. While many individuals with LQTS have persistent prolongation of the QT interval, some individuals do not always show the QT prolongation; in these individuals, the QT interval may prolong with the administration of certain medications.

All forms of the long QT syndrome involve an abnormal repolarization of the heart. The abnormal repolarization causes differences in the "refractoriness" of the myocytes. After-depolarizations (which occur more commonly in LQTS) can be propagated to neighboring cells due to the differences in the refractory periods, leading to re-entrant ventricular arrhythmias.

It is believed that the so-called early after-depolarizations (EADs) that are seen in LQTS are due to re-opening of L-type calcium channels during the plateau phase of the cardiac action potential. Since adrenergic stimulation can increase the activity of these channels, this is an explanation for why the risk of sudden death in individuals with LQTS is increased during increased adrenergic states (ie exercise, excitement) -- especially since repolarization is impaired. Normally during adrenergic states, repolarizing currents will also be enhanced to shorten the action potential. In the absence of this shortening and the presence of increased L-type calcium current, EADs may arise.

The so-called delayed after-depolarizations (DADs) are thought to be due to an increased Ca^{2+} filling of the sarcoplasmic reticulum. This overload may cause spontaneous Ca^{2+} release during repolarization, causing the released Ca^{2+} to exit the cell through the $3Na^+/Ca^{2+}$-exchanger which results in a net depolarizing current.

The diagnosis of LQTS is not easy since 2.5% of the healthy population have prolonged QT interval, and 10% of LQTS patients have a normal QT interval.

There are two treatment options in individuals with LQTS: arrhythmia prevention, and arrhythmia termination.

Arrhythmia suppression involves the use of medications or surgical procedures that attack the underlying cause of the arrhythmias associated with LQTS. Since the cause of arrhythmias in LQTS is after depolarizations, and these after depolarizations are increased in states of adrenergic stimulation, steps can be taken to blunt adrenergic stimulation in these individuals.

Arrhythmia termination involves stopping a life-threatening arrhythmia once it has already occurred. The only effective form of arrhythmia termination in individuals with LQTS is placement of an implantable cardioverter-defibrillator (ICD). (11) ICD are commonly used in patients with syncopes despite beta blocker therapy, and in patients who have experienced a cardiac arrest.

It should be noted that there are also some miscellaneous causes of QT prolongation such as anorexia nervosa, hypothyroidism, HIV infection, and myocardial infarction. (6,7,10)

"An abnormal repolarization of the heart" is what happens to the heart during and after myocardial capture has occurred, or also stated as "when the CEW drives the heart". A myocardial infarction (an interruption of the blood supply) happens during myocardial capture (the heart suddenly pumps out more blood than is coming in), heart attacks, ischemia (restriction in blood supply), oxygen shortage in the blood, and runners syndrome, all of which was completely covered in the author's preceding report. (1)

Dr. Zian Tseng's submission to the Braidwood inquiry (next page) shows an elongated QT cardiac cycle graph. (9)

In the authors opinion, the CEW's by forcefully driving cardiac muscle at it's maximum potential replicates the damages that this syndrome causes. The Long QT Syndrome (LQTS) does fully explain the sudden in-custody deaths after being shot by a CEW. Additional side effects as documented have added complications.

Currently, the only long term solution for LQTS damage is by an implantable cardioverter-defibrillator (ICD). This information can potentially put Taser International board member Mark W. Kroll in a conflict of interest situation. He sits as a board of director for other medical institutions and companies. He has numerous papers, a large number of patents, inventions, and co-written a book on ICD's. This statement is just identifying the issues, it is not intended to defame or be slanderous.

64 · CONDUCTIVE ELECTRONIC WEAPONS AND THEIR FAULTS

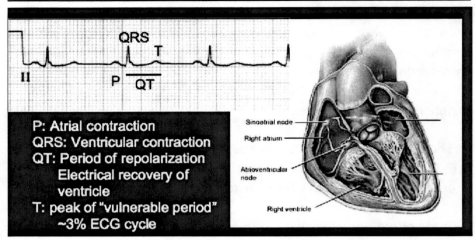

Graph provided by Dr. Zian Tseng. (9)

Next addendum, Dr. Sloane has also shown a Magnetic Resonance Image the brain after a CEW probe has punctured the right parietal. (3) This image shows the electromagnetic field created by the MRI machine being induced into the CEW probe and emitting into the brain, with the brain tissue displaying the pattern. This is a similar effect as described on page 23 Fig **C**. of the radiating pattern of electrons from the probe, also on next page.

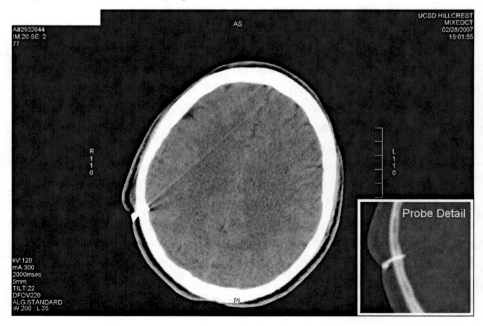

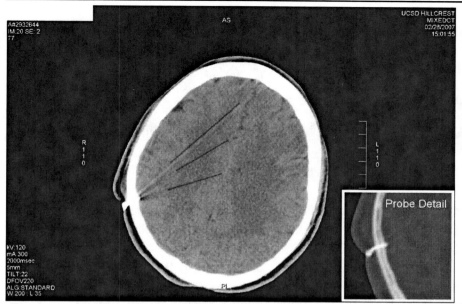

Electromagnetic radiation pattern induced into the probe from the MRI machine.

Next addendum, Stinger Systems, the competitor to Taser International, has provided information relating to a software problem that can effect the accuracy of the information provided by the weapon.

"An arbitrary date can be mistakenly set on Taser's data capture and if that date is incorrectly put in, then the firing information is also inaccurate." (8) Below is a screen capture taken from Taser International material provided to the Braidwood Inquiry. (4)

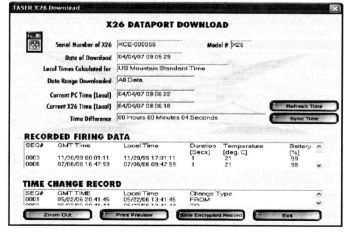

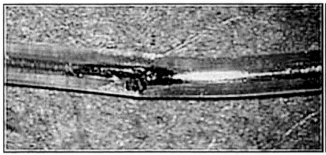 

Image Courtesy of Stun Wire Shot Technology©

It has been reported that the wires used in the Taser International 35ft. cartridges have a weakness in their high voltage limits. The insulating material used is ETFE (Tefzel™) which has a rating of 80kv/mm. However when the insulation is reduced in thichness to .17mm, (the coating as applied to the cartridge wires) the voltage insulating rating drops to 27kv/mm. The Taser Devices charge the wires to 50kv.

The problem identified is that when the darts are shot into a moving subject, or a subject with heavy or baggy clothing in long range shots, for whatever reason if the circuit is not complete, that's when the problem occurs.

The stun gun charges the wires to full capacity, but because the voltage does not discharge through the subject, this maintained high voltage in the wires causes a breakdown of the insulation. If the wires cross, there is a possibility of the wires shorting out. Taser International developed the 25ft cartridge to overcome this problem, where as the shorter range allows for more penetrating power by the probes, and more chance of a complete circuit.

Information supplied by Stun Wire Shot Technology©

Chapter 5 Update January 15, 2009

Taser International claims their device does not cause venous gases. (1) However testing procedures and faults have been described (3), electrolysis is a known and verified electrochemical reaction.

It can be stated that the human body can handle small amounts of air or gases within the blood stream. Any patient receiving intravenous (IV) fluids will receive and witness a few bubbles in the tubing fluid flow. It is taught that less than 25cc's of air continuously should not cause problems to a patient, however more than 25cc's of air can cause an embolism.

Small amounts of air do not produce symptoms because the air is broken up and absorbed from the circulation. Although classical teaching states that more than 5mL/kg of air (IV) is required for significant injury (including shock and cardiac arrest), patient complications secondary to as little as 20mL of air (the length of an unprimed IV infusion set) have been reported. Further, as little as 0.5mL of air in the left anterior descending coronary artery has been shown to lead to ventricular fibrillation. (2)

Embolism

A blood clot or other solid mass, as well as an air bubble, can be delivered into the circulation through an IV and end up blocking a vessel; this is called embolism. Peripheral IVs have a low risk of embolism, since large solid masses cannot travel through a narrow catheter, and it is nearly impossible to inject air through a peripheral IV at a dangerous rate. The risk is greater with a central IV.

Air bubbles of less than 30 milliliters are thought to dissolve into the circulation harmlessly. Small volumes do not result in readily detectable symptoms, but ongoing studies hypothesize that these "micro-bubbles" may have some adverse effects. A larger amount of air, if delivered all at once, can cause life-threatening damage to pulmonary circulation, or, if extremely large (3-8 milliliters per kilogram of body weight), can stop the heart.

One reason veins are preferred over arteries for intravascular administration is because the flow will pass through the lungs before passing through the body. Air bubbles can leave the blood through the lungs. A patient with a heart defect causing a right-to-left shunt is vulnerable to embolism from smaller amounts of air. Fatality by air embolism is vanishingly rare, in part because it is also difficult to diagnose. (4)

Background

Venous air embolism (VAE), the entry of gas into the peripheral or central vasculature, can occur secondary to iatrogenic complications, trauma, and even certain recreational activities. Although many occurrences of VAE are believed to go unreported because they are asymptomatic, entrapment of large quantities of intravascular gas can lead to severe neurologic injury, cardiovascular collapse, and even death. The factors that determine the subsequent morbidity and mortality in VAE include the rate of air entrainment, the volume of air introduced, and the position of the patient at the time of the embolism.

Gas emboli are usually composed of air, but they can also occur with medically used gases such as carbon dioxide, nitrous oxide, and nitrogen. Although very small volumes of air can lead to severe sequelae, it is generally accepted that more than 50 mL of air can cause hypotension and dysrhythmias and more than 300 mL of air can be lethal.

Pathophysiology

VAE results when a pressure gradient develops that favors the ingress of air into the venous system. Upon entry into the venous system, air is transported to the right atrium and ventricle. From there, it has the potential to continue on to the pulmonary arteries where it may cause interference with gas exchange, cardiac arrhythmias, pulmonary hypertension, and even cardiac failure and arrest. A large bolus of air entering the venous system can cause an air lock in the right atrium and ventricle, leading to outflow obstruction, decreased pulmonary venous return, and subsequent decreased left ventricular preload and cardiac output.

Intermediate amounts of air collect in the pulmonary circulation and produce a pulmonary vascular injury manifested by precapillary and postcapillary pulmonary vasoconstriction, pulmonary hypertension, endothelial injury, and permeability pulmonary edema. Subsequent ventilation-perfusion mismatch can cause right to left shunting and increased arterial hypoxia and hypercapnia.

Small amounts of air do not produce symptoms because the air is broken up and absorbed from the circulation. Although classical teaching states that more than 5 mL/kg of air (IV) is required for significant injury (including shock and cardiac arrest), patient complications secondary to as little as 20 mL of air (the length of an unprimed IV infusion set) have been reported. Further, as little as 0.5 mL of air in the left anterior descending coronary artery has been shown to lead to ventricular fibrillation.

The pathogenesis of pulmonary endothelial injury may have components of platelet-fibrin thrombi from the right ventricle, cytokine release, neutrophil, platelet, and complement activation at the microvascular air-blood interface, and injury mediated by lipid peroxidation and oxygen radicals.

Mortality / Morbidity

VAE is associated with significant morbidity and mortality. Morbidity can include lung injury, neurologic injury, cardiovascular ischemic injury, and ultimately cardiopulmonary collapse and arrest. Symptomatic VAE following CV catheterization has a mortality rate as high as 30%.

History

Signs and symptoms of VAE usually develop immediately following embolization. Severity of signs and symptoms are related to the degree of air entry into the body. The diagnosis can be a difficult one to make because of its similarity in presentation to pulmonary embolism (thromboembolic) and/or hypovolemic shock. Subsequently, physicians must maintain a high index of suspicion for this disease given the appropriate clinical scenario. The following historical events should be considered in taking a patient's history for suspected VAE:

- Recent surgery, especially neurosurgical, cardiovascular, or orthopedic procedures
- Blunt and penetrating trauma to the face, neck, chest, and / or abdomen
- Recent invasive procedures such as central venous catheterization or pressurized infusion of fluids, blood, or contrast
- Patients with indwelling central venous catheters
- Decompression injuries/sickness
- Orogenital sex during pregnancy or hydrogen peroxide ingestion (rare)

- Symptoms may include the following:
  - Dyspnea
  - Chest pain
  - Agitation or disorientation

Physical

Physical examination may reveal the following signs:

- Tachycardia
- Tachypnea
- Cyanosis
- Altered level of consciousness
- Hypotension
- Cardiac "mill wheel" murmur - A loud, churning, machinery like murmur heard over the precordium (a late sign)

- Sudden loss of consciousness followed by convulsion in an intubated patient on positive-pressure ventilation
- Circulatory shock or sudden death (patients with severe VAE)

A fourth cause of VAE is positive-pressure ventilation, which can occur during mechanical ventilation and SCUBA diving.

Finally, blunt and penetrating trauma to the chest, abdomen, neck, and face can lead to the entry of air and ultimately to VAE. (5)

All of the above symptoms are in whole or in part associated with people that have been shot by a CEW device. How venous gases and acidosis are created, and the mechanism of spiking is covered in detail in the diagram on page 57. The mechanism of the 30 minute increase and 30 minute decrease is also explained. The people that have died shortly after a CEW discharge would not have their venous gases measured until the autopsy, which happens after this 60 minute time line has expired. No accurate readings would be found from the victims under the current autopsy system.

During the course of compiling information it was decided that there should be no attempt to rewrite news articles. They are all highly informative, direct to the point, are included verbatim and credited!

This author contacted Mr. Robert Anglen of the Arizona Republic Newspaper as he covered the story of Taser International and the US Air Force testing the device. His article is reprinted here. (6)

Taser tied to 'independent' study that backs stun gun
The Arizona Republic: May 21, 2005

Taser International was deeply involved in a Department of Defense study that company officials touted to police departments and investors as "independent" proof of the stun gun's safety, according to government documents and e-mails obtained by *The Arizona Republic* and interviews with military officials.

This information is surfacing at a time when the U.S. Securities and Exchange Commission and the Arizona attorney general are pursuing inquiries into safety claims that the Scottsdale firm has made.

The stun guns are being used by more than 7,000 law enforcement agencies in the United States, but a series of deaths and injuries associated with the devices have raised safety concerns.

E-mails that military officials exchanged also reveal for the first time that they asked Taser to tone down public statements about the study. In addition, they urged the company to commission an independent study rather than rely on the Defense study.

The Air Force conducted the study for the Defense Department to assess the risks and effectiveness of Tasers so the military could decide whether to buy them.

Since October, Taser officials have contended that the company had no involvement in the Defense study, which helped fuel a sharp rise in the company's stock price last year.

Bulk of research

But information obtained by *The Republic* shows that Taser officials not only participated in three panels to determine the scope of the study, analyze data and review findings, it also provided the bulk of research material used in the study.

"Were they (Taser) totally disconnected (from the study)? The answer is no. They were not disconnected," said Larry Farlow, a spokesman for the Air Force Research Laboratory in Texas that oversaw the study.

Taser critics - civil rights lawyers, human rights activists and government officials - contend that there is insufficient evidence to support the company's assertions that the stun gun is safe. They have called for independent research.

Taser has repeatedly characterized research that its own employees or consultants helped conduct or write as independent. The company has also paid training fees and given valuable stock options to police officers involved in decisions to purchase the stun guns.

In an interview earlier this month, Steve Tuttle, Taser's vice president of communications, maintained the company's position that the Defense Department study was independent. He acknowledged that Taser employees had some involvement in the study but insisted that that did not influence the findings.

Taser officials have described the Defense research as "a major independent safety study." But Air Force researchers said the study was not meant to be a comprehensive review of stun-gun science or safety, and they made no findings on the device's safety.

Touting findings early on

Taser trumpeted results of the study long before the actual report came out on April 1. In an October news release, Taser Chief Executive Officer Rick Smith said, "This comprehensive independent study further supports the safety of Taser" and "reaffirms the lifesaving value of Taser technology."

That announcement had an immediate impact on Taser stock: It shot up 60 percent during the next month. Taser executives and board members sold 1.28 million shares for $68 million in November.

Since then, the stock has dropped dramatically as a series of deaths caused cities nationwide to reconsider purchases of Tasers and to delay deployments.

An ongoing investigation by *The Republic* has found that medical examiners have cited Tasers in 15 deaths across the country. They called it a cause of death in three cases, a contributing factor in nine cases and said the stun gun couldn't be ruled out as a cause of death in three cases.

Taser maintains that its stun guns have never caused a death.

Taser involvement

When the Defense Department first released its study, it made no mention of who was involved in the study.

Another version obtained by *The Republic* shows that Taser's CEO, director of technical services, general counsel, medical director, chief instructor, electrical engineer and vice president of communications were involved in various panels over five months.

The report also shows that companies doing business with Taser, including General Dynamics, were heavily involved in the study and, along with Taser executives, sat on a final "Independent External Review Panel" to examine all the findings.

Farlow, the spokesman for the Air Force Research Laboratory, said his office, not Taser, made the decision to strike the names from the final report in order to protect the privacy of researchers and scientists.

A separate panel of medical and scientific experts that did not include Taser employees wrote the final report.

Tuttle, the Taser spokesman, said the company's involvement does not minimize the report's significance or its independence.

"This was all pre-planning stuff," he said. "We didn't do the study itself." He added that government rules require manufacturers to be involved in such reviews of their products. "If you are going to do a study of Milk Duds . . . you are going to have to talk to the (makers) of Milk Duds."

But, according to the Air Force, Taser provided most of the data used in the study, which was supposed to look at the "effectiveness" of Tasers in order to provide guidance for officials in charge of purchasing non-lethal weapons.

Information gaps

Although researchers determined the stun guns were "generally effective for their intended use," researchers found significant "data gaps" in the information Taser provided, Farlow said.

Chief among those gaps: enough information to determine whether Tasers can cause seizures or induce ventricular fibrillation, the sudden irregular heartbeat characterized by a heart attack.

In addition, Taser apparently did not provide some information about injuries involving the stun gun. For example, researchers said in the study that "no reports were identified that describe bone fractures resulting from the rapid induction of strong muscle contraction" caused by the stun gun.

At the time that Taser officials were sitting on the panel, they had already been served legal notice that a Maricopa County sheriff's deputy was going to sue the company over a fractured back that he reportedly suffered when shocked with a Taser during a training exercise.

Former Deputy Samuel Powers was the first to file a product liability lawsuit against Taser; his case is scheduled to go to trial in June. A doctor hired by Taser last year concluded that a one-second burst from a Taser was responsible for Powers' injury.

Since then, several police officers from departments across the country have come forward with allegations of bone fractures that they blame on Taser shocks.

The study concluded that Tasers may cause several unintended side effects, "albeit with estimated low probabilities of occurrence." It also said the need to "rely on a database of case reports compiled by manufacturers also generates uncertainty in the results."

Farlow pointed out that the Defense study made no conclusions about the stun gun's safety.

When asked about Taser's characterization of the research as a "major, independent safety study," Farlow said: "The simple answer is consider the source. . . . The press and public relations folks are doing their jobs."

E-mail correspondence

Despite the fact that the Air Force lab's study made no findings on safety, the government officials who commissioned the study allowed Taser to issue a news release saying that the Defense Department considered "Tasers generally safe and effective."

E-mails show that although these officials were concerned about Taser's characterization of the study, their desire to support Taser prevailed.

"I've expressed my personal view to (Taser) that the company might want to take a different approach to their (public affairs) efforts" and "i.e., tone it down," wrote Capt. Daniel McSweeney, spokesman for the Joint Non-Lethal Weapons Directorate, a Pentagon office that recommended purchasing Tasers for the armed services.

"My opinion is that they probably want to commission an independent (human effects) study, in which a variety of stakeholders participate," McSweeney said in a January e-mail from his office in Quantico, Va. "To settle this issue once and for all."

Dave DuBay, a Taser vice president, confirmed that McSweeney asked the company to temper its statements. He said McSweeney felt Taser is sometimes "too passionate in defense" of its stun guns. DuBay also confirmed that McSweeney asked Taser to commission its own independent study.

But DuBay said the government's study was independent and questioned whether the public would perceive a Taser-sponsored study to be independent.

Despite McSweeney's concerns, he still recommended backing Taser.

McSweeney's rationale

"My rationale is that Taser is, in effect, some kind of partner to us, since we purchase and field their systems," he wrote in the same e-mail. "Not supporting them can hurt us in the public's eye."

At issue in the e-mails were requests from Taser asking the government to put out a news release declaring the stun guns safe.

The e-mails were written after reports in the New York Times and other media raised questions over Taser's claims about the Defense study and if researchers actually found the stun guns safe.

In an interview this week, McSweeney confirmed that he told Taser officials they should "tone it down" and conduct their own independent study.

"I was referencing not just to the (study) but other things I have been privy to," he said, adding that Taser has been at the center of several controversial issues. "Given the ongoing questions regarding the health effects of Taser, it would behoove Taser to do an independent study."

McSweeney acknowledged that the Defense study was not comprehensive but called it an "excellent first step" and said that more studies are under way. He said that non-lethal weapons are needed in military zones and that the study served "an urgent need" by providing a foundation for the Defense Department.

Taser International disputes CBC / Radio Canada testing as flawed

CBC News: Dec 10, 2008

Taser International is responding to moves by several Canadian police agencies to pull some stun guns off the streets for testing following a recent CBC/Radio-Canada investigation into the devices, saying the U.S. laboratory tests commissioned by the CBC and Radio-Canada are "flawed."

In written releases to news agencies, the Arizona-based company says the CBC investigation made scientific errors by failing to spark-test the weapons before firing them, which the company recommends to police officers.

However, Taser International's testing protocol used by CBC doesn't call for a spark test before conducting measurements. In addition, the CBC test found some units still delivering higher current after the equivalent of a spark test.

Taser International is also criticizing the CBC tests over the way the tests replicated electricity moving through a human body, which is measured in ohms.

The CBC tests followed the company's protocol and written instructions from Max Nerham, Taser International's vice-president of research and development. Nerheim advised using a resistance of 250 ohms when testing the Taser. After seeing the CBC test results, the company is saying that resistance should have been 600 ohms.

The tests, conducted by the U.S.-based lab National Technical Systems, found about 10 per cent produced more electrical current than the weapon's specifications. The malfunctioning Tasers were manufactured before 2005.

This week, municipal police forces in B.C. joined the RCMP in suspending the use of all Tasers bought before Jan. 1, 2006.

Solicitor General John van Dongen said Tuesday the weapons will be tested to ensure that the electrical currents generated are consistent with the manufacturer's specifications.

"We are establishing a provincial standard for both testing and calibration so that we know that all of the equipment that's in service meets the required specifications," he said.

Van Dongen said the government has taken the action in the interest of public and officer safety.

Other police forces make similar moves

Meanwhile, other police jurisdictions across the country are also planning to test the older Tasers in the wake of the CBC/Radio-Canada investigation. Ottawa police are removing some of their Tasers from service and testing others.

The Ottawa police tactical and explosives unit has 32 Tasers, nine of which were manufactured before 2005, confirmed Staff Sgt. Mike Maloney, who is in charge of the unit. All nine older Tasers will be retired and replaced with newer devices.

The rest, all newer units, will be re-tested, even though they met standards in earlier tests, Maloney said. "We're just going to make sure we have the data to show that, yes, these Tasers are working properly."

The Winnipeg Police Service is also planning to remove older Tasers from service.

Police will decide in due course whether older X26 models of the stun guns will be tested and returned to service or new equipment will replace the old ones, a police spokesperson said Wednesday.

Of 191 Tasers in the Winnipeg police arsenal, 41 have serial numbers indicating they are older X26 models — made before 2005.

On Wednesday, Nova Scotia's justice department ordered police forces in the province to stop using the older Tasers.

Quebec was the first province to respond to the CBC/Radio-Canada testing.

On Dec. 5, the day after the test results were made public, Quebec's public security minister immediately ordered all police departments to take Tasers older than 2005 off the streets.

Police use of Tasers has generated intense public concern after Polish immigrant Robert Dziekanski died at Vancouver International Airport more than a year ago. An RCMP officer hit him with a Taser shortly before his death.

The Braidwood Inquiry, which has been looking at the use of Tasers and circumstances surrounding Dziekanski's death, will continue to examine the use of the weapons by municipal police, SkyTrain police, sheriffs and corrections officials in the second phase of the public hearing.

Statement by Taser International spokesman Peter Holran:

"It is regrettable that false allegations based on scientifically flawed data can create such uncertainty. Taser International stands behind the quality and safety of its products and is prepared to provide the assistance and information necessary to allay any concerns.

Taser International welcomes proper testing of its devices and has provided its factory test protocols to test laboratories in Canada so police agencies can avoid the scientific errors made by the CBC. Using proper test protocols will ensure that going forward decisions are not based on scientifically flawed data such as was presented in the CBC report.

According to the data provided by CBC from its test, ALL 41 Taser devices tested produced energy outputs consistent with the expected outputs published by Taser International. It also is apparent from the data that four devices each produced what can only be explained as an anomaly in peak current during one of six firings — the first firing of each device tested at 250 ohms — most likely the result of the testers failing to spark test the Taser device before the test — a requirement made to all officers and agencies in the training as a check for proper function and to condition electronic components."

The CBC / Radio Canada testing of the X26™ devices does not reveal any new results that have not been published before. Dr. Robert Walter of the Cook County Trauma Unit was given a copy of this authors report, he had an opportunity to proofread it and supplied some testing results and background information.

Taser Internationals main reason for disagreeing with the CBC report is that the test resistance used was incorrect. The average resistance load was 30 to 250 ohms (10), whereas Taser recommends 600 ohms be used. Taser's testing found the internal resistance between 400 to 800 ohms, as tested on humans and pigs, nothing about the dogs. (8,9)

Taser International is incorrect with their resistance measurements, and was not using established medical information. As covered on page 18, Philips, the company that manufacturers Defibrillators states in their literature that the human bodies resistance is between 25 to 180 ohms (7,15). Once the Taser probes penetrate the skin, (the only organ that was designed to resist electricity), the electricity is freely able to flow through the conductive blood supply.

As previously stated, if the area around the probe has been irrigated with blood from exercise, (such as running away from the police) that again would only lower that internal resistance. On January 30, 2008 Taser International Rick Smith states in a CBC / National interview that "the pig heart is not like the human heart", so Taser International has discarded it's own testing data.

All of Taser Internationals stun products should be retested using the 25 to 180 ohms resistance as covered in Philips literature (7,15), there's no other way to state it!

Note: Philips frequently changes it web site around, if these locations don't have the information on body resistance then contact them through their customer service for the latest information.

The following is taken from Cameron Ward Website:

In British Columbia, we still allow police officers to investigate themselves after they kill someone. The eminent former judge and legal scholar, the Hon. Roger Salhany, QC, recently described this approach as "a bad, if not intolerable, idea" (Report of the Taman Commission of Inquiry, Manitoba, 2008).

The police investigators generally spend a long time, not trying to make a case that will stick, but trying to create an airtight defence that will absolve police of responsibility for the death. Then, unlike conventional cases where the investigators recommend specific charges, they deliver a "neutral" report to Crown Counsel that contains no recommendations at all. Little wonder then that, as far as I can tell,

"No BC police officer has ever been prosecuted for a civilian death resulting from the intentional application of force in the 150 year history of this province."

In this case, the RCMP conducted a lengthy investigation into the incident that involved four of its members, even taking the time and trouble to fly to Poland to reportedly try to dig up dirt on the unfortunate victim. It will be interesting to see when and whether the investigators even interviewed the RCMP members involved, or whether they followed the usual practice of "debriefing" them and inviting them to contact legal counsel and submit a statement in writing at their leisure.... (11)

Halifax Chronicle Herald: December 17, 2008

On Friday (Dec 12), the B.C. Attorney General's report on the Dziekanski incident found criminal charges were not warranted against the officers involved. Given the loose nature of the RCMP's Taser policy, that's not surprising. But the report went further, calling the Mounties' actions – which included five Taserings and the aggressive physical restraint of Mr. Dziekanski – "reasonable and necessary."

The RCMP were quick to say they've made changes to their Taser use rules since the event. But they also admitted those new rules would not have made any difference in the officers' actions on the day that Mr. Dziekanski died.

Police forces in Canada continue to seem to be in denial about the public's well-grounded concerns about police use of Tasers.

The RCMP and the office of the B.C. Attorney General are dangerously out of touch. If their rules say that what happened was OK, then their rules need to be changed, before another innocent person dies at police hands. (12)

Vancouver Sun: December 13, 2008

New details emerged Friday as the Crown announced no charges will be laid against the four RCMP officers involved in the death of a Polish man who was Tasered on Oct. 14 last year at Vancouver International Airport.

Stan Lowe, speaking on behalf of the Criminal Justice Branch, which oversees charges and prosecutions in B.C., revealed for the first time that Robert Dziekanski had received five shocks from a Taser over 31 seconds, not two as initially announced by the RCMP.

RCMP Supt. Wayne Rideout, the team leader of an investigation conducted by members of the Integrated Homicide Investigation Team, said Friday an RCMP spokesman had initially issued incorrect information that the Taser had been deployed only twice.

The mistake was made in the rush to provide the media information immediately after the incident, he said.

"The RCMP spokesman conveyed the information he had been provided from one of the officers present at the airport. That officer did not himself deploy the conducted energy weapon," Rideout told reporters at a news conference.

Once it was realized incorrect information was released, police could not correct it because a criminal investigation was under way and police did not want influence the investigation, he said.

And while police initially called to the scene thought Dziekanski was intoxicated, the Crown revealed Friday that no drugs or alcohol were found in his system, and that he had an unopened bottle of Polish vodka with him.

Dziekanski's mother, Zofia Cisowski, was upset when told by officials on Thursday that no charges would be laid against the four officers, her lawyer, Walter Kosteckyj, said Friday.

"Her initial reaction was, 'How can you tell me a Taser was deployed five times on my son and it isn't excessive?'" the lawyer recalled.

She also questioned why officers "jumping" on her son to restrain and handcuff him wasn't considered excessive force, he said.

"They [the RCMP] said they don't have any legal responsibility," Kosteckyj said, but added that his client plans to file a civil lawsuit against police.

"They just mishandled the entire situation," said the lawyer and former Mountie.

He pointed out that Dziekanski, seconds before he was first jolted with the Taser, told police in Polish: "Have you lost your mind? What are you doing?"

"They may have followed procedure," Kosteckyj said of the police actions. "If the procedure was flawed, that's not an issue for the criminal justice system, but it is an issue for the Braidwood commission."

The Braidwood commission of inquiry into the incident and the use of Tasers by police resumes Jan. 19 in Vancouver, presided over by retired judge Thomas Braidwood.

The Crown also revealed Friday that the Taser shocks did not cause the cardiac arrest that led to Dziekanski's death, but that the actions of police in restraining and handcuffing the man were found to be a contributing factor. (This requires more specific details to released, but has not been).

(Known contributing factors; 4 officers held down Mr. Dziekanski. He appeared to be clubbed at least three times with a collapsible baton in the video, the CEW was used 5 times, three in probe mode, two in stun mode. And one officer had his knee on his throat!)

Page 38 covers the risks of placing the prisoner in the prone position, which does not apply in this case. In cases of the recovery position, Mr. Dziekanski was on his right side the whole time. An issue with venous gases becomes a concern because when placed on the right side, the human body is much more vulnerable than on the left side. "Immediate first aid for a large amount of air (or or other gas) that enters the blood stream is to turn the patient onto their left side with the head elevated to trap the air bubble in the upper right atrium of the heart. (2)

Lowe told the news conference that the available evidence from the police investigation "falls short" of the branch's charge-approval standard, which is a substantial likelihood of conviction.

He said the decision was reviewed by three levels of prosecution administration, including Robert Gillen, the assistant deputy attorney-general of B.C.

The charges considered were assault, assault with a weapon and manslaughter, Lowe said.

"We were all horrified by what happened," Attorney-General Wally Oppal said Friday.

But Oppal pointed out that the law entitles police to use reasonable force, and the conclusion reached by prosecutors was that the force used was justified under the circumstances.

"The public should realize the process is not over yet," the attorney-general said, adding that the Braidwood commission will hold the officers and the RCMP accountable for their actions.

Dziekanski, 40, had arrived in Vancouver from Poland after 21 hours of travelling. He exhibited bizarre, aggressive behavior at the airport, including throwing furniture and a computer, before he was encountered by police, who responded after several people called 911.

Witnesses told police the man was "freaking out, drunk and did not speak English," Lowe explained during a news conference Friday.

"The officers attempted to talk to Mr. Dziekanski and communicate with him with hand signals for several seconds," he added. "He momentarily calmed down and dropped his arms to his side."

Then Dziekanski became annoyed and frustrated, threw up his arms, moved to his right, grabbed a stapler and held it out in his hand, Lowe said. The officers then backed away from Dziekanski and deployed Taser for the first time.

Cause of death was listed as "sudden death following restraint," Lowe said.

The officers involved were identified only by their rank and last names: Constables Millington, Bentley and Rundel, and Cpl. Robinson.

Millington was the officer with the Taser. It was revealed for the first time that the Taser was deployed three times in probe mode -- in which two electrical probes are fired at a person and conduct an electrical shock -- and twice in stun mode.

Lowe said the evidence shows Millington initially deployed the Taser in probe mode but the device appeared to malfunction because it was making a "clacking" sound, indicating the probes were not making proper contact, resulting in an incomplete electrical circuit. *(See pages 25 - 31, and 66 for reasons of poor electrical contact and variable performance)*

After Dziekanski was on the ground and continued to struggle, Millington deployed the Taser twice more in stun mode, he said.

It took about 30 seconds after the last Taser shock for officers to restrain and handcuff Dziekanski, Lowe said.

A use-of-force expert from a police force outside the RCMP reviewed the police investigation and concluded the officers' actions were consistent with RCMP policy and training, he said.

The RCMP also disclosed Friday that two of the officers involved in the incident had been transferred to undisclosed locations after a number of "ugly" taunting incidents. Police wouldn't elaborate.

One officer remained in B.C., working on "other functions" -- not front-line duties -- and another has been suspended for an unrelated incident, said RCMP Assistant Commissioner Al Macintyre, who is in charge of criminal operations in B.C.

He said the four Mounties will cooperate with the Braidwood inquiry, which has been delayed awaiting the charge-approval decision.

The incident has been difficult for the RCMP and the officers involved, said Macintyre, who acknowledged the grief it has caused Dziekanski's mother.

"The RCMP is committed to learning as much as possible from this incident and making adjustments to its policies and practices where needed," Macintyre said.

"Since this incident, the RCMP has made a number of changes to its conducted energy weapon [Taser] policies, training, practices and reporting requirements, and we are certainly open to making further improvements," he said.

He said changes already made include: restricting the use of the weapons to incidents involving threats to officer or public safety; requiring RCMP officers to be re-certified annually on the use

CONDUCTIVE ELECTRONIC WEAPONS AND THEIR FAULTS

...ducted energy weapons; testing of the weapons themselves; enhanced use-of-force reporting; and analysis of reporting on conducted energy weapon usage.

"There are some who believe the conducted energy weapon should no longer be used by police. There is obviously a lot of emotion around this issue," Macintyre said, adding that policy decisions should be based on facts and scientific data, not emotion.

The RCMP has initiated independent electronic testing of a sampling of Tasers, including voltage variance, Macintyre said.

He said the airport death was still being investigated by the B.C. coroner service, which will hold an inquest, and the RCMP Public Complaints Commission.

MLA Mike Farnworth, the NDP public safety critic, said the Braidwood inquiry needs to answer many questions the public has about the incident and the reasoning behind the charge-assessment decision.

"There are a lot of outstanding questions the public needs answers to," Farnworth said. "I think the public will want to know why [charges were not laid] -- the logic and reasoning behind it."

How far did the RCMP have to drive?

The RCMP officers are not based in the terminal, How far did they have to travel?

The RCMP station at YVR International Airport is not located in the terminal. It was located in a building off site. Once a call is made, It takes one minute on a good day for the RCMP to drive from their station to the terminal, and under five to walk / run. On this night, The three officers waited outside the terminal for three to four minutes (14) for the forth officer with the CEW to arrive from the Richmond detachment.

The officers were clearly notified by several members of the public that Mr. Dziekanski did not speak English, yet as the video shows, the officers made no attempt at communication or any attempt to diffuse the situation! The several seconds of hand gestures as claimed by one of the officers looks like "Ok buddy, up against the wall". Anyone who watched the video can reach their own conclusions that the officers only intention was to see a new toy in action. There is no reason why the officers could not wait or diffused the situation for another ten minutes so an officer who understood polish (and was near by) could arrive and help, as reported by the local tv media.

The officer with the CEW drive took 6 minutes to arrive on site after the call at 1:21am. (14)

The airport is covered in video cameras, it is not known why a video of the officers outside the terminal was not shown in court. Rumors persisting that the officers waited outside for up to 30 minutes waiting for a CEW equipped officer to arrive started from live news reports later that day. Actual police route shown below.

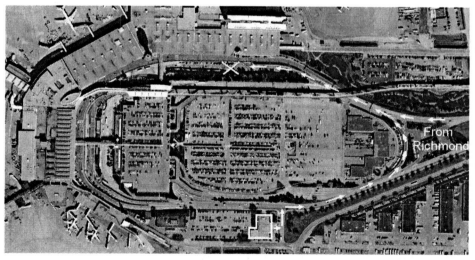

The BC Coroners service was contacted 5 times during the course of this investigation. The author submitted this report and asked several questions. Although the receptionist was helpful, The BC Coroners Service did not reply to any questions asked, or even returned messages. The questions were about testing time lines and methods.

The Canadian Press: December 17, 2008

OTTAWA–Opposition parties want the RCMP to stop using Taser stun guns after the force refused to reclassify the weapons to restrict use.

The Liberals and NDP say the Mounties missed a parliamentary committee's deadline Monday to categorize the 50,000-volt electronic devices as impact weapons.

Reclassifying Tasers would limit use to situations where a person assaults police or the public, or poses a serious threat of harm or death.

An RCMP spokesperson said no progress report was sent to the committee because Parliament isn't sitting. (it is currently prorogued) The RCMP says reclassifying the Taser as suggested could threaten police and public safety. (13)

YVR Remodels the arrival area

YVR International Airport remodels the international arrivals area for the 2010 Olympics. The insulating glass has been removed, but that door on the left still remains.

Chapter 6 Update April 30, 2009

The US Army has concluded that the Taser can cause seizures and ventricular fibrillation. Robert Anglen from the Arizona Republic was contacted and provided this story. It is reprinted in full.

Study raises concerns over Tasers' safety

The Arizona Republic: Feb 13, 2006

A study measuring electric shocks from a Taser stun gun found that it was 39 times more powerful than the manufacturer claimed, raising new questions about the weapon's safety.

The study, published last month in the peer-reviewed Journal of the National Academy of Forensic Engineers, concluded that the shocks are powerful enough to cause fatal heart rhythms. It is one of the few scientific studies of Taser's electric jolt in which the company did not participate.

"The findings show the energy delivered by the weapon to be considerably understated by the manufacturer," the Journal study said. "These findings place the weapon well into the lethal category."

Officials with Scottsdale-based Taser International Inc. condemned the findings, saying they are exaggerated, erroneous and "beyond the laws of physics."

They pointed to a test conducted last week in response to the Journal article. A lab hired by Taser found that the weapon produced power that was significantly less than what the Journal study found and met all specifications.

Taser contends that the author of the Journal study, electrical engineer James Ruggieri, does not have the technical expertise to make conclusions about stun guns. Taser is suing Ruggieri for defamation over his claims in a presentation and testimony in a wrongful-death case last year that Tasers can cause fatal heart rhythms.

In a separate finding, the Army also concluded last year that Tasers could cause ventricular fibrillation, the irregular heart rhythm characteristic of a heart attack.

A memorandum from the Aberdeen Proving Grounds in Maryland, where the Army develops, tests and evaluates weapons, said, "Seizures and ventricular fibrillation can be induced by the electric current."

At issue was whether soldiers should be shocked with the stun guns during training exercises, as Taser recommends.

The Army's occupational health sciences director determined that Taser is an effective weapon but added in the February 2005 memo that "the practice of using these weapons on U.S. Army military and civilian forces in training is not recommended, given the potential risks."

Taser for years has maintained that its stun guns have never caused a death or serious injury. Company officials say the guns save lives, reduce injury and save millions of dollars in legal costs because they prevent deadly confrontations.

But since 1999, more than 167 people have died after police Taser strikes in the United States and Canada. Of those, medical examiners have cited Tasers in 27 deaths, saying that they were a cause of death in five cases, a contributing factor in 17 cases and could not be ruled out in five cases.

Several law enforcement agencies have filed lawsuits accusing Taser of misleading them about the stun gun's safety and claim that the company failed to conduct adequate tests before selling the weapon. Some police departments have delayed or halted Taser purchases because of safety concerns.

Taser denies these claims and says its record of safety is bolstered by dozens of medical and university studies and by the company's experts.

Law enforcement officials and testing experts agree that there is no widely accepted standard for measuring Tasers. Studies have shown various results.

In May, for example, an international testing laboratory hired by Canadian authorities initially reported that two stun guns were significantly more powerful than the manufacturer specified. The guns also fired at different levels of power.

The stun guns were used on a man who died after being shocked by Vancouver, British Columbia, police in 2004.

Taser challenged the test last week, and the laboratory backed off its results. Officials with the lab, Intertek ETL Semko, said testing protocols provided by the police differed from those of the stun-gun manufacturer. As a result, Intertek said the tests could not be relied upon.

Bruce Brown, deputy commissioner of a British Columbia agency investigating the police role in the Vancouver death, said his agency wants to enlist Canada's National Police Research Center to conduct a rigorous study of the stun gun's power.

"We've sent people to the moon, so there has got to be a way to come up with a peer-reviewed (standard)," he said.

The 50,000-volt Taser works by shooting two darts up to 25 feet. The darts are connected to wires that deliver a burst of electricity that is designed to instantly immobilize a suspect. The gun also can be used as a handheld device, without the darts, by touching two metal probes directly against a person's body in what police call a "drive stun."

The shock from a Taser is measured in electric pulses. Tasers typically used by police deliver 15 to 19 pulses a second in a five-second interval, although the gun will continue firing without interruption as long as the trigger is held down.

Tasers operate at 50,000 volts, but Taser says the stun guns do not pose an electrical safety risk because the pulse's current is too low and its duration too short to affect internal organs, including the heart.

Ruggieri's study found that the Taser's pulse was more powerful and longer than the gun's specifications indicate. Ruggieri studied a Taser M-18, which is nearly identical to the Taser M-26 used by police except it has less power.

Taser specifies that the M-18 produces 10 pulses a second at 1.76 watts per pulse. Ruggieri said his tests showed the Taser produced 14 pulses a second at 50 watts per pulse.

Ruggieri said it took him months of research to conduct and complete the tests.

He said he relied on Taser's research and previous stun-gun studies to create a verifiable methodology for testing the Taser.

His findings are based on how electric current penetrates the body. When established electrical standards were applied to the stun gun's electrical discharge, Ruggieri said the current could be fatal. He said measurements of the electric current showed that, according to electric safety standards, the gun had a 50 percent risk of causing ventricular fibrillation.

Taser Vice President Steve Tuttle called the claim "ludicrous" and said it is "clearly refuted by the fact that well over 100,000 human volunteers have been exposed to the Taser discharge without fatality."

Taser maintains that skin tissue blocks electric current and is equivalent to 1,000 ohms of resistance.

But Ruggieri said skin tissue breaks down as electricity is applied, decreasing resistance and increasing the impact of the shocks on the human body.

"This creates a runaway effect of increasing current with decreasing resistance," Ruggieri said.

An independent electrical engineer who reviewed the Journal study at the request of *The Arizona Republic* said Ruggieri's conclusions were credible and based on scientific principles.

Robert Nabours, who has degrees in electrical engineering from Stanford and the University of Arizona, said scientific and medical evidence support Ruggieri's claims that skin tissue breaks down when subjected to electric pulses. Among the evidence are findings from Harvard and Massachusetts Institute of Technology doctors.

Ruggieri focused on the Taser in its "drive stun" mode. He said measurements of the current found that the power was about 39 times greater than the manufacturer's specifications. Taking into account the lowered resistance of skin tissue, Ruggieri said the stun gun generated 704 watts of power as opposed to 18 watts.

Ruggieri contends that one of Taser's main claims of safety, that the duration of the electric pulse is too short to cause injury, could not be proven. He said his tests of the current showed that duration of the pulse also increases as resistance drops.

The lab hired by Taser, Exponent of Phoenix, could not replicate Ruggieri's results. Exponent, which has offices throughout the country, is a consulting firm that employs scientific and engineering experts who, like members of the National Academy of Forensic Engineers, often serve as expert witnesses in court cases.

Exponent electrical engineer Ashish Arora said Ruggieri reported 17 times more power than the Taser he tested. Arora said that in his tests, the power of the stun gun measured at or below specifications.

Arora said the pulses Ruggieri measured could also not be verified, even when resistance was dropped. He said that caused concern.

He said he would have expected some similarity in the results. But he said the tests results "were completely different."

There were differences between Exponent's and Ruggieri's tests, both involving how the gun was charged and how the current was measured.

Ruggieri said he used a battery specified by the manufacturer to mirror a real-world setting. He changed the battery after each jolt to ensure that the power did not degenerate. Exponent used a power supply to charge the battery.

Ruggieri said a power source could limit the amount of power going into the gun in a way that a battery would not.

Ruggieri also measured the output using two high-voltage meters attached to each of the Taser probes, which he said gave more-accurate readings.

Exponent used a single meter. Arora said the single probe and battery wouldn't change the results.

Taser has repeatedly attacked Ruggieri's credibility since he made a presentation critical of the stun guns to the American Academy of Forensic Sciences in February 2005. Taser claimed his presentation was based on "junk science" and "propaganda" and that his conclusions have been disputed by numerous government, university and medical studies.

Some of Ruggieri's claims were independently verified, including his assertion that Taser had misapplied Underwriters Laboratories standards in suggesting the stun gun could not cause ventricular fibrillation.

Taser sued Ruggieri in November, several months after he announced the Journal findings at an engineering conference in Chicago.

In a news release last year, Taser described Ruggieri as a high school dropout with no medical training.

Ruggieri said he left high school to attend college in New York. He later obtained a master's degree in computer science from the University of Phoenix.

Ruggieri's resume shows that he is a professional engineer with licenses in five states. He said he has investigated electrical accidents for federal agencies and helped write electrical safety standards for top electrical laboratories and commissions.

Taser officials challenged the academy journal, calling it an "obscure bulletin," saying none of the peer reviewers was qualified to assess the findings.

"That unfortunately allowed Mr. Ruggieri to utilize inappropriate science and flawed mathematics in attempts to support his unsupportable conclusions," Taser's Tuttle said.

Journal Editor Marvin Specter said the academy is affiliated with the National Society of Professional Engineers and is made up of experts in several engineering disciplines.

The Journal lists a technical review committee for Ruggieri's study that includes 20 engineers, including one well-known Taser consultant. The reviewers' identities are confidential and have not been released, Specter said.

Specter said Ruggieri's paper went through a rigorous peer-review process before being published in the biannual journal.

In an interview last week, Ruggieri said Taser has launched personal attacks to distract from the real issue.

"This isn't about me. It's about the findings, the study," he said. (2)

Review of the postmortem examination.

The postmortem was finally made available to the author on April 26, 2009. The Petechiae, and the probe impact locations will be covered. (6)

The Petechiae as stated by Dr. Lee is consistent with resuscitation attempts. Petechiae is simply broken blood vessels, broken from too much blood pressure. The weakest link in any chain will break first, and that link is under the eye lids. Petechiae is associated with asphyxiation, however, that alone won't cause Petechiae, strangulation or constriction of the neck is required. The blood must be able to pump into the head, with the constriction not allowing the blood to exit freely. The extra pressure causes the blood vessels to burst, and that is Petechiae.

There is an excellent article called "Asphyxial Deaths and Petechiae: A Review. (3) This article goes in depth and should be read in detail.

Most of the article can be quoted, but will be left out so it can be read and understood in it's entirety.

A second report titled, "Death from Law Enforcement neck holds" (American Journal of Forensics Medicine and Pathology, Sept. 1982) (4) covers how the officers choice of restraint could have contributed to his death, however after reading those two articles, it is not possible for Mr. Dziekanski's heart to be pumping any kind of meaningful blood flow at that point. The snoring (Agonal breathing) he was making at this point means the cardiac arrest or fibrillation had already happened! Also petechiae would have been more apparent in more locations if Mr. Dziekanski's body was still in full working order. Both those reports are available on the Charly D Miller web site. (5)

The Charly D Miller web site contains many more documented problems with associated police restraining procedures, unsafe restraint, in-custody deaths, asphyxia, Tasers, etc. (5)

The postmortem revealed that only one probe managed to penetrate the skin, while the other only left punctate wounds (page 5 under commentary). Punctate means spotted, marked, with points or punctures, or depressions. As Mr. Lee could not find the point of entry for the second lower probe, this means the second wire was not permanently connected and the electrical connection only worked intermittently.

The assumption that the stun gun will work as per the manufacturers directions assumes the probes will both be properly seated into the targets skin and muscle, and no acts of god or intermittent contacts are encountered. According to the postmortem, what has happened is the top probe seated properly and made a good connection, while the lower probe did not. This lead the Taser to surge, or pulse pump out much more voltage and current than normal. Pulsing a signal is nothing new, it is actually one method used to get more power out of a circuit, but for a device that can effect the body, it is a very unwanted condition.

As previously covered in Taser International literature, there are two parts to the operation of the device. The first part of the circuit closing is called the "Arc Phase" where the electricity first begins to arc over and burn through the wires insulation creating a plasma arc, this allows the electricity to flow. This start up is the only time where the device is at 50,000 volts or above, in normal advertised operation.

The device can not maintain this power level and the voltage quickly drops down to the 1200 to 5000 under a load, also as previously covered in CEW report 3. The second part is what Taser International calls the "Neuromuscular Incapacitation" phase. The 1200 to 5000 volt level, is not suppose to lead the heart, but is suppose to freeze the subject. However the intermitten wire connection will cause a change in the operation of the Taser device that is not on their web site or in the operations manuals of the devices. It should be known to their engineers that an intermittent connection will cause a pulsing scenario with the power amplification circuit.

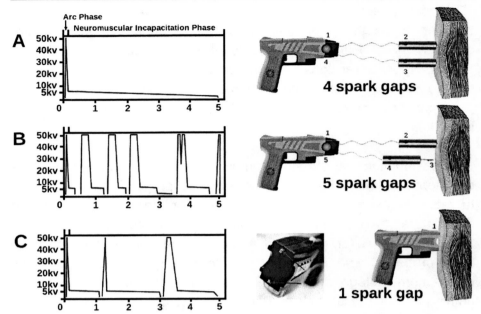

A) Normal advertised operation. The Arc phase burns away the wire insulation (2+3) to make electrical contact with the probes, and electricity travels into the subject.

B) An intermittent contact with the lower probe causes surging and power spikes. (3) Since the insulation on the wires is already burnt away the peak power periods is extended as the electricity is not required to remove the wires insulation (2+4). Any kind of continuity break (1-5) allows the circuit to recharge, and will pulse on the next available re-connection. The continuity breaks can be of any duration, this graph show a few of the possible variation over a 5 second period. More breaks means more pulses. This type of voltage amplification circuit generally operates in the 10 to 50 kilohertz range, so time for full recharge is nearly instantaneous. Once the Arc is made, it can exist on much less power, and transfer more.

C) In Drive Stun mode, the electricity does not enter the body, it passes on the outside of the skin causing burn damage and pain (pain compliance). Contact can be broken, by clothing, or hair. As the electricity did not penetrate the skin, this method is not likely to have killed him directly, but contributed by adding pain and strain to his body.

With every break in the connection, that will cause the device to recharge to its maximum potential, and at the next re-connection, will pump the Arc phase all over again. The circuitry is suppose to deliver one short arc, and pulse 15 to 19 times per second, for five seconds. However, if the connection is broken say once per second, then it will deliver 6 Arc pulses per 5 second shot. If the connection is broken three times per second, the the device will deliver 16 Arc pulses per the 5 seconds.

An analogy can be made to turning on an old style light bulb. Light bulbs almost always burn out shortly after you turned them on. The reason for this is the inrush of current was always greater at first turn on, and that causes the connections to fail. That same inrush of current is also applicable to the stun guns. The Arc phase can be considered as that inrush of current.

The stun gun discharge makes a loud crack the first time as the arc is created because of the electrical current jumping over the built in spark gaps. After the first Arc, there is no more insulation to burn away, so the electricity has an easier time jumping the gap. If the intermittent connection is brief enough, then there may not be any noticeable sound effects, but power fluctuations flowing could still be huge. For a 5 second shot, there may be several interruptions, but no second loud crack. As already covered, there is no feedback that the operator can use to judge when this happens, officer Millington also testified to this fact.

Spark gaps are switches (7), and depending on the design do not necessarily reduce power levels. In high voltage designs, it allows the electricity to build up until the voltage is high enough to be able to jump over the gap. The wider the gap is, the more electricity is building up. All faults of the spark gap design were covered in the CEW report 3 and updates.

Looking at Taser International's patent 6,999,295, there is a spark gap internally on the circuit board, between the high voltage circuit and step-up transformer. As the casing is not hermetically sealed, the difference of conditions between completely dry air, and high

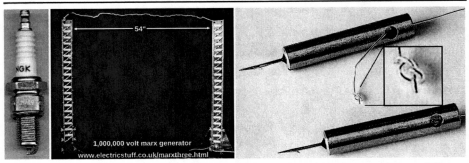

Spark gap designs, a spark plug, a 1,000,000 volt generator, and the probes enameled wires.

atmospheric water vapor can alter the voltage the device puts out before any pulse reaches the cartridge.

A simple analogy is ask anyone who drove an old car with a point distributor system. On warm sunny days the car ran fine, but on cold rainy days the car had an attitude until it warmed up! If you washed the cars engine, you had to be careful not to get any moisture under the distributor cap, or else the car wouldn't turn over at all! After a while, you would have to do maintenance and clean off the points due to it getting dirty and fouled. The reason for this is the arcing path burns the pollutants in the atmosphere, and it electroplates onto the point contacts. All stun guns will need periodic cleaning for reliable service!

Note: The Author could not find out if the older model devices are capable of recording the repeated arcing cycle scenario of the output section. Looking at some of their patents 6,999,295, 7,234,262, 7,457,096, 7,570,476, this does not appear to be the case. The newer X3™ model is advertised as having this capability, however it is unclear where the device tests the signal. Signals leaving the device before the spark gaps or the returning ground? If it only tests from the output pulse before it enters the cartridge, then there is little expectation to see any distortion of the signal, its just a direct feedback. Once the pulse goes through any one of the spark gaps, the movement of the wires can break the connection, and reshape the pulse into something else than what was transmitted. Once the circuit is broken, the power amplifier circuitry will recharge, sending a stronger pulse out at the next connection.

EXHIBIT

FINAL REPORT
REPORT OF POSTMORTEM EXAMINATION

#76

VANCOUVER HOSPITAL
Forensic Pathology
855 West 12th Avenue
Vancouver, B.C. V5Z 1M9
Tel: (604) 875-4024 Fax: (604) 875-4768

DATE OF AUTOPSY October 16, 2007

COMMENCED AT 09:00 Hours

CORONER Owen Court

CORONER'S NUMBER 07-270-1054 ✓

AUTOPSY # 07-2512

DECEASED'S NAME DZIEKANSKI, Robert

SEX Male

DATE OF DEATH October 14, 2007

AGE 40

PRINCIPAL PATHOLOGICAL FINDINGS:

1. No significant injuries.
2. Fatty liver, atrophy of the cerebellar vermis, dilated heart.
3. Negative toxicologic examination.
4. No atherosclerosis.

RECEIVED
JAN 29 2008
MINISTRY OF SOLICITOR GENERAL
B.C. CORONERS SERVICE
VANCOUVER METRO REGION

PART 1. PRINCIPAL CAUSE OF DEATH:

a. _____Sudden Death During Restraint_____

b. _____due to or as a consequence of_____

c. _____due to or as a consequence of_____

PART 2. CONTRIBUTORY FACTORS:

a. _____Chronic Alcoholism_____

b. _____

C. Lee, MD, FRCPC
Forensic Pathologist

DZIEKANSKI, Robert 2
07-2512

POSTMORTEM EXAMINATION:

The autopsy was conducted in the mortuary of the Vancouver General Hospital on Tuesday, October 16, 2007 commencing at 09:00 hours.

Attending the autopsy were Sergeant Christiansen of the Richmond RCMP and Constable Hoivik of IHIT. Pons Paez was the Pathology Attendant.

IDENTIFICATION:

Identification was by the Coroner's Form B. There was a tag on the body bag bearing the deceased's name and the Coroner's File #07-270-1054.

EXTERNAL EXAMINATION:

State of the Body

Temperature:	Cold
Rigor Mortis:	Full
Postmortem Lividity:	Posterior surface of body, fixed
Decomposition:	Nil

External Features

Sex:	Male
Age:	Consistent with documented age of 40 years
Length:	177 cm
Weight:	86 kg
Build & Nutrition:	Average build. Well nourished
Hair:	Dark brown
Eye Colour:	Hazel
Pupils:	0.5 cm, equal
Conjunctivae:	Slightly congested, rare petechiae in the left lower eyelid
Nose:	Intact
Teeth:	Natural upper and lower dentition
External Genitalia:	Unremarkable male
Skin Features:	Extensive irregular scars are on the posterior left upper arm and the left side and back suggestive of burns. A 5.0 cm faint linear scar is on the anterior right thigh.

EVIDENCE OF THERAPY:

Oral airway, endotracheal tube, ECG pads, defibrillator pads, intravenous catheter in the left antecubital fossa, needle puncture marks on the right arm covered by bandages.

DZIEKANSKI, Robert 3
07-2512

Injuries associated with resuscitative attempts include bilateral anterior 2nd to 6th rib fractures and a fractured sternum. Scattered small abrasions are on the central chest.

EVIDENCE OF INJURY:

Head and Neck

A 2.5 x 1.0 cm abrasion is above the left eyebrow. A 3.0 x 2.0 cm abrasion is on the right cheek. Internally, there are no skull or brain injuries, and a layerwise dissection of the neck reveals no injuries.

Thorax and Abdomen

A pair of punctate abrasions 2.0 cm apart is on the central chest. The lower abrasion is somewhat darkened around the edges. A 4.0 x 2.0 cm brown contusion is on the lower right chest. A punctate dried red abrasion is on the lower right abdomen. There are no internal injuries.

Extremities

A 3.0 x 2.0 cm red contusion is on the left upper arm. Multiple round and oval red contusions, approximately 1.5 to 2.0 cm in diameter, are on the posterior left forearm. A 3.0 cm transverse dried linear abrasion is on the lateral left wrist, and a 5.0 cm linear faint red contusion is on the medial left wrist. A 1.0 cm abrasion is on the back of the left hand near the base of the index finger. A 0.5 cm abrasion is on the lateral right forearm. A 6.0 cm linear red contusion is on the medial right wrist. Incising both wrists reveal scattered subcutaneous hemorrhages in both wrists.

Several linear scabbed abrasions are on the lateral left thigh. A 2.5 x 1.5 cm red contusion is on the left knee. A 1.0 cm dried red abrasion is just below the left knee. A 3.0 cm red contusion is on the medial left ankle. A 0.5 cm dried red abrasion is on the medial right shin.

INTERNAL EXAMINATION:

Head and Neck

Scalp:	Unremarkable
Skull:	Intact
Brain Weight:	1190 gm
Brain - External:	Cerebral hemispheres are symmetrical and mildly atrophic, but otherwise unremarkable. Brainstem and cerebellum are unremarkable. The cerebral arteries contain no atherosclerosis.
Brain - Sections:	Brain is cut fresh. Ventricles are symmetrical and of normal size. No localized lesions. The brainstem is unremarkable. The cerebellum shows atrophy of the cerebellar vermis.

DZIEKANSKI, Robert   4
07-2512

Vertebral Column:	Intact
Larynx:	Intact with no mucosal lesions
Hyoid Bone:	Intact
Thyroid:	Unremarkable
Soft Tissues of Neck:	Unremarkable

Thorax

Rib Cage:	See Evidence of Therapy
Pleural Cavities:	No significant effusions or adhesions
Mediastinum:	No tumours or lymphadenopathy
Esophagus:	Unremarkable
Trachea & Bronchi:	Clear of foreign material
Pulmonary Vessels:	No thromboemboli
Lungs:	Right 1030 gm and left 990 gm; congested and edematous cut surface with no localized areas of consolidation.
Pericardium:	No significant effusions or adhesions
Heart Weight:	370 gm
Heart:	Right ventricle is 0.4 cm in thickness, left is 1.3 cm in thickness. Valves are unremarkable in structure and circumference and without surface vegetations or scarring. The ventricles are dilated, and the myocardium appears somewhat soft but shows no scars or acute infarction.
Coronary Arteries:	Right dominant circulation; all widely patent
Aorta & Major Arteries:	No atherosclerosis

Abdominal Cavity

Peritoneal Cavity:	No significant effusion or adhesions
Stomach & Duodenum:	No ulcers or other mucosal lesions; lumen contains minimal contents
Intestines:	Contain minimal contents, and contains no evidence of drug packets. No obstruction or perforation is noted.
Appendix:	Present
Liver:	1800 gm; soft, greasy yellow/tan cut surface with no localized lesions
Gallbladder:	No calculi
Bile Ducts:	Normal calibre
Spleen:	130 gm; unremarkable cut surface
Pancreas:	Unremarkable
Adrenals:	Unremarkable
Kidneys:	Right weighs 150 gm and left weighs 160 gm; smooth tan/brown cortical surfaces with no localized lesions
Urinary Bladder:	Contains no urine. Mucosa unremarkable

Musculoskeletal System/Extremities

DZIEKANSKI, Robert 5
07-2512

No evidence of acute or chronic bone or joint deformity.

TOXICOLOGY:

Samples of blood and vitreous fluid are submitted to the Provincial Toxicology Centre for toxicological analysis.

The report of that analysis (PTC#07-1671) indicates no ethanol or other drugs detected. No glucose or ketones are detected in the vitreous fluid.

MICROSCOPIC DESCRIPTION:

Kidney: No significant histopathology.

Liver: Severe macrovesicular steatosis; mild to moderate chronic inflammation in the portal tracts, without associated hepatocyte necrosis.

Heart: Mildly increased, patchy interstitial and perivascular fibrosis; focal areas of acute interstitial hemorrhage; mild fatty infiltration in the right ventricle, without fibrosis or inflammation.

Lung: Vascular congestion, rare fat embolism. No thromboemboli identified.

Brain: The cerebellum shows moderate to severe atrophy, with loss of Purkinje cells and proliferation of Bergmann's glia. The remainder of the brain is unremarkable.

COMMENTARY:

The circumstances surrounding the death were provided to me by the Coroner's Form B as well as medical records provided by Citizenship and Immigration Canada, and a video of the events leading to death. The autopsy showed only minor, superficial injuries, including bruising around the wrists consistent with the application of handcuffs. There were no neck injuries. The sternum and ribs were fractured, but these are consistent with injuries occurring during resuscitation attempts. A darkened punctuate abrasion on the central chest is consistent with an electrode from a Taser. The other electrode mark is not apparent, but a couple of punctuate abrasions are present on the chest and abdomen, and one of these might be the other Taser mark. Rare petechial hemorrhages are noted in the left eyelid, but these could have occurred as a result of the resuscitation attempts. The autopsy also showed the presence of a severely fatty liver, atrophy of the cerebellar vermis, and a dilated cardiomyopathy. These findings are consistent with chronic alcoholism. The remainder of the autopsy was unremarkable. Postmortem toxicological examination was unremarkable. Sudden death following restraint has been described in the forensic literature, but the cause and mechanism is not well understood. It has been associated with virtually all forms of

DZIEKANSKI, Robert
07-2512

physical restraint. It usually involves men who are combative and acting bizarrely. As a result, these cases often involve law enforcement personnel. However, cases have also involved medical personnel, and occasionally ordinary citizens. In many of these cases, the deceased was intoxicated with drugs such as cocaine or other stimulants. The current belief is that the drugs cause the agitated or excited delirium that result in the bizarre behaviour, and the subsequent death following restraint. Typically, the autopsy shows minimal findings.

This case differs from the cases typically described in the literature in that the toxicology examination shows no drugs present. Furthermore, no medical condition that may cause delirium was identified. This corresponds with the video which shows the decedent to be agitated, but he did not appear delirious. However, the absence of a definite anatomic cause of death is typical of the sudden deaths following restraint that is described in the literature. His dilated cardiomyopathy would have put him at an increased risk for development of an arrhythmia and sudden death, but probably would not have caused death by itself. The added stress of the physical restraint along with the decreased ability to breathe as a result of being pinned in the prone position may have been enough to elicit a fatal arrhythmia. The presence of signs of chronic alcohol abuse does raise a possibility that he was suffering from alcohol withdrawal, which may partly explain his agitation. It is likely a combination of these and other contributory factors that lead to his death. Therefore, the cause of death is best described as sudden death following restraint.

Charles Lee, MD, FRCPC
Forensic Pathologist

CL/pd

Chapter 7 Update July 27, 2009

A review of Mr. Dziekanski's Toxicology.

The following was compiled from several emails to the Braidwood inquiry.

The Toxicology report states no Glucose or Ketone's in the Liver, which is consistent with the last stages of operation before Liver failure matching alcohol abuse. Impaired cognition is also one side effect of Liver failure in conjunction with the other problems found in the postmortem.

The author believed at one point that Water Intoxication could be part of the problem. The news media previously reported that he was given three bottles of water by security staff to drink during his time there (reported by Global TV) (update Mar 28, 2008 CBC News, now revised to 5 to 6 glasses of water). It is the author's speculation that, If that was the case, it is very possible that he was already in a state of minor water intoxication when the incident happened. In the video, his apparent heavy sweating is one method the body uses to remove excess fluid. Heavy sweating gives the suggestion to the individual that he should drink more, leading to this confusion. The body does not give any warning that there is too much water intake until the Electrolyte level fluctuations cause complications. Water Intoxication such as this cause the victim to feel nausea, irritability, pain, and eventually delirium.

The full CBSA report, released under the freedom of information act, on pages 15, 19 & 21 it states that only one glass of water was given to him when he motioned to the officers for a glass of water. The court testimony of witnesses stated that there were no visible signs of confusion or strange behavior for 24 hours before, but less than one hour after drinking the water his behavior changed.

It could be he was under a state of minor confusion the whole time he was lost in the YVR baggage and arrivals area.

A side effect of Liver failure is Acidosis, but as Acidosis was not recorded on the Postmortem or the Toxicology report, it would not appear to be at that stage yet. As Liver failure is not reversible, a transplant was in his eventual future.

Now a note about that last paragraph, and this has been stated by this author before. Since autopsies are not done at the time of death (October 14), the acidosis, and lactic acid would only increase with the process of Rigor Mortise, and no usable readings be gathered. His conditions was full Rigor Mortis at the time of the autopsy, October 16.

What is Lactic Acidosis?

Lactic acidosis is a condition caused by the buildup of lactic acid in the body. It leads to acidification of the blood (acidosis), and is considered a distinct form of metabolic acidosis. Tissue hypoxia and hypoperfusion force cells to breakdown glucose anaerobically. The result is lactic acid formation. Therefore elevated lactic acid with clinical signs and symptoms is indicative of tissue hypoxia, hypoperfusion and possible damage. Lactic acidosis is characterized by lactate levels >5mmol/L and serum pH <7.35.

Many cells in the body normally burn glucose to form water and carbon dioxide. This is a two-step process. First, glucose is broken down into pyruvate by mean of glycolysis. Then the mitochondria, by means of the Krebs cycle, oxidize pyruvate into water and carbon dioxide, using oxygen. When there is a lack of oxygen in the blood, then the mitochondria cannot burn all the pyruvate produced by glycolysis, and pyruvate accumulates in the cell. This accumulation cannot be tolerated by the cell which converts pyruvate to lactate and evacuates it into the blood, hence causing lactic acidosis.

The signs of lactic acidosis are deep and rapid breathing, vomiting, and abdominal pain. Lactic acidosis may be caused by diabetic ketoacidosis or liver or kidney disease, as well as some forms of medication (most notably the anti-diabetic drug phenformin). Some anti-HIV drugs (antiretrovirals) warn doctors in their prescribing information to regularly watch for symptoms of lactic acidosis caused by mitochondrial toxicity. There is some doubt as to whether lactic acid truly produces 'acidosis', rather it is thought to be from ATP conversion to ADP.

Lactic acidosis is an underlying process of rigor mortis. Tissue in the muscles of the deceased resort to anaerobic metabolism in the absence of oxygen and significant amounts of lactic acid are released into the muscle tissue. This along with the loss of ATP causes the muscles to grow stiff. (1)

What is Rigor Mortis?

Rigor Mortis is a condition that affects a human body after death. Rigor mortis caused the muscles in the body to become rigid. It starts to happen three to four hours after death. Twelve hours after death, the muscles have become fully rigid. The condition usually lasts until 36 hours after death. At that time, the muscles start to relax.

After death, respiration in organisms ceases to occur, depleting the corpse of oxygen used in the making of ATP. ATP is no longer provided to operate the SERCA pumps in the membrane of the sarcoplasmic reticulum, which pump calcium ions into the terminal cisternae. This causes calcium ions to diffuse from the area of higher concentration (in the terminal cisternae and extracellular fluid) to an area of lower concentration (in the sarcomere), binding with troponin and allowing for crossbridging to occur between myosin and actin proteins.

The body may only move, after rigor mortis has set in, if put into a below freezing temperature, and forty-five minutes later switched to a higher temperature. Repeat this cycle to get the muscles moving. (2)

The following is taken from an email to the Braidwood Inquiry

A review of the probe impact locations.

Using Gray's Anatomy as a guide to where the probes landed, the upper probe landed in the central chest below the sternum, in the Rectus Abdominus muscle group. This location was also seen on the postmortem picture shown to Dr. Tsang on the live video stream. Judge Braidwood made a comment to the Taser International lawyer to be more careful with the evidence a moment later.

This location directly connects with the Falciform ligament, a remnant of the persons umbilical chord. It also directly connects to the thoracic diaphragm, which separates the upper respiratory organs from the lower gastronomic organs, and assists in breathing and respiration. Any direct Neuromuscular Incapacitation of this group would restrict breathing, and could assist in the modulation of his vocal scream as heard in the Pritchard video. The Falciform Ligament also directly connects to the left lobe of the Liver which has multiple vein and artery connections to the heart.

But before the probe penetrates to that depth it penetrates through the Peritoneum. The Peritoneum is the serous membrane that forms the lining of the abdominal cavity. The peritoneum both supports the abdominal organs and serves as a conduit for their blood supply, lymph vessels and nerves. This leaves multiple pathways for the electrical discharge to take, both external through the skin and internal through the organs.

This may contradict Dr. Tsang's testimony, but a review of pictures of the internal ribcage (If any exist) for punctures would be in order.

Note: I have not received feedback from the Inquiry if this was done? The body had been cremated by this point.

PROVINCIAL TOXICOLOGY CENTRE EXHIBIT #116
Phone: (604) 707-2714

B.C. CORONERS SERVICE

Coroner COURT, OWEN
#6 1818 CORNWALL AVE.
VANCOUVER, BC V6J 1C7

TOXICOLOGY REPORT

Dr. LEE, C.
VANCOUVER GENERAL HOSPITAL

Name of Deceased: DZIEKANSKI, ROBERT [Alias: DZIEKANSKSI, ROBERT]

PTC No: 2007-1671

Coroner Case No: 2007-00270-1054   Autopsy No: 07-2512

The Provincial Toxicology Centre received on 17 October 2007 the following specimens:
BLOOD (GREY TOP TUBE): 1 BLOOD (RED TOP TUBE): 2 BRAIN (TUB): 1 KIDNEY (TUB): 1
LIVER: 1 VITREOUS (GREY TOP TUBE): 1

The results are as follows: (* indicates new result)

Specimen	Source	Test	Result
BLOOD (GREY TOP TUBE)	FEMORAL	ETHYL ALCOHOL	Not Detected
BLOOD (GREY TOP TUBE)	FEMORAL	OPIATE (SCREEN)	Not Detected
BLOOD (GREY TOP TUBE)	FEMORAL	COCAINE & METABOLITES (SCREEN)	Not Detected
BLOOD (GREY TOP TUBE)	FEMORAL	AMPHETAMINES/GEN. DRUG SCREEN	No Drugs Detected
BLOOD (GREY TOP TUBE)	FEMORAL	ACIDIC DRUG SCREEN	No Drugs Detected
BLOOD (GREY TOP TUBE)	FEMORAL	BASIC DRUG SCREEN	No Drugs Detected
BLOOD (GREY TOP TUBE)	FEMORAL	GC-MS (SCREEN)	No Drugs Detected
BLOOD (GREY TOP TUBE)	FEMORAL	BENZODIAZEPINE (SCREEN)	No Drugs Detected
BLOOD (GREY TOP TUBE)	FEMORAL	TRICYCLIC ANTIDEPRES (SCREEN)	No Drugs Detected
BLOOD (GREY TOP TUBE)	FEMORAL	NEUROLEPTIC DRUGS (SCREEN)	No Drugs Detected
BLOOD (GREY TOP TUBE)	FEMORAL	PCP	Not Detected
*LIVER		ETHYL ALCOHOL	Not Detected
*LIVER		OPIATE (SCREEN)	Not Detected
*LIVER		CANNABINOIDS (SCREEN)	Not Detected
*LIVER		COCAINE & METABOLITES (SCREEN)	Not Detected
*LIVER		AMPHETAMINES/GEN. DRUG SCREEN	Not Detected
*LIVER		ACIDIC DRUG SCREEN	No Drugs Detected
*LIVER		BASIC DRUG SCREEN	No Drugs Detected
*LIVER		GC-MS (SCREEN)	No Drugs Detected
VITREOUS (GREY TOP TUBE)		ETHYL ALCOHOL	Not Detected
VITREOUS (GREY TOP TUBE)		GLUCOSE (BY DIPSTICK)	Not Detected
VITREOUS (GREY TOP TUBE)		KETONES (BY DIPSTICK)	Not Detected

NO ALCOHOL, PRESCRIBED MEDICATION OR ILLICIT DRUGS DETECTED.
ADD-ON
AS PER COMMUNICATION WITH THE CORONER/PATHOLOGIST ALTERNATIVE SPECIMEN (LIVER) WAS ALSO ANALYZED.

Walter Martz, Ph.D., Forensic Toxicologist

Report issued on 15-November-2007   Exhibits will be discarded after 13-February-2008

Answers to statements made during the trial.

Q: Taser International repeatedly claims "high voltage electricity only flows between the probes"

A: The pulsed electricity from the device will enter the body and exit the body from the probes, but the path it chooses to take in-between is not a straight line. It will follow the conductive path of least resistance. When it can not find a direct path, it will search the conductive fluids and tissues for another route, such as in the picture below and the following page.

A series of pictures using Tesla coils demonstrates that electricity will indeed follow the path of least resistance. A Tesla Coil oscillates to create a high frequency voltage. As the device discharges into the atmosphere, the direct path it takes in the atmosphere goes through electrolysis. That path suddenly becomes non-conductive as the hydrogen and oxygen atoms split, the electricity is then no longer able to travel that route and chooses the next available path. These pictures show that process. This would also happen within the body of a person being shot with a CEW. The CEW's pulse is a repeating cycle, the number per second depends on the model of the device. If the electricity only went through one muscle to get to the ground probe, then only one muscle will be affected. But since the device freezes the whole body (as demonstrated in the company advertisements) then the electricity must be directly going through multiple paths to do that.

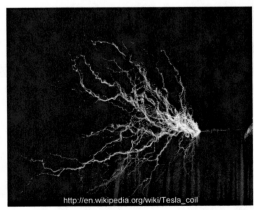

Electricity that is searching a conductive medium (air) without a ground.

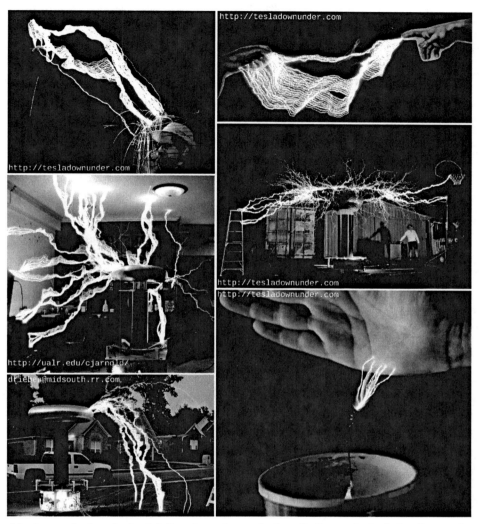

Various shots of electricity choosing what path it wants to take through the conductive atmosphere to find a ground.

Why is there an Ozone smell after an electrical discharge or lightning? The answer is covered by Electrolysis. The voltage splits the atmospheric water vapor into Hydrogen and Oxygen. The Hydrogen will want to take off, but the Oxygen molecules will want to bond into O_2, and then into O_3 groups for about 30 minutes. O_3 is Ozone.

J. Patrick Reilly from the Applied Physics Laboratory of Johns Hopkins University, gave a power point presentation to the Inquiry on may 5, 2008. *The transcript was reviewed late.*

The output of the M26™ and X26™ models are not correct from past documentation. His testing found the M26™ now delivers 7000 volts under a load, such as probe mode embedded into a person. The power output and this model is only suppose to deliver 1200 volts. That's a 5.8 times increase of power.

According to the X26™ documentation, it is suppose to deliver 1500 volts, but Mr. Reilly said it's 1300 volts under load. He also stated the X26™ builds up a 50,000 volt charge then the charge drops to 25,000 volts as the probes dangle in the air *(discharge through the air - author)*

He later references several cases, one where a police office was severally injured from a CEW discharge, and a case where a persons shoulder was severally shattered from a muscle contraction, with current discharge lower than a stun gun could deliver.

Finally he states that (in probes that don't penetrate the skin) wet skin is no more dangerous than dry skin to a CEW arc discharge. The reason he gives is the skin is automatically transformed into a good conductor by the discharge? The water is instantly transformed by the discharge plasma leaving only dry skin remaining.

The author laughed when he read this because splitting the water molecules this rapidly by the plasma will cause burning of the skin tissues. He does not go into details when preexisting conditions are present such as broken skin or blood is present.

Refer to the picture on the previous page of a persons hand getting hit by an electrical discharge, this is not to be done without expecting pain! A burn pattern can be seen on the lower edge of his hand, in the (dry) skin to the right of the discharge. The web site has a better resolution image.

Chapter 8 Update April 4, 2010

Taser International Inc. sues Canadian government
The Arizona Republic: Aug 15, 2009

Taser International Inc. filed a lawsuit Friday in Canada blasting a government report that prompted severe limitations on how and when law-enforcement officers in British Columbia can use stun guns.

Officials with the Scottsdale-based manufacturer called the Braidwood Inquiry biased and asked the Supreme Court of British Columbia to quash all of its findings and declare those involved in compiling evidence derelict.

"We provided . . . more than 170 studies, periodicals (and) reports with respect to the safety of the device and use-of-force questions," David Neave, an attorney for Taser in Canada, said Friday. "All of that information clearly indicates that when the device is used properly there is not cardiac effect. For reasons unknown to us, that information did not wind its way into the report."

The 18-month-long Braidwood Inquiry, headed by retired Judge Thomas Braidwood, concluded in July that Tasers can cause death.

In his 556-page report, Braidwood criticized law enforcement for putting the stun gun on the street with little or no independent testing and recommended restricting use of Tasers. Within hours, the head of public safety in British Columbia adopted all 19 of Braidwood's recommendations, including a ban on Tasers in non-criminal situations or where there is not an imminent threat of bodily harm.

A spokesman for the Braidwood Inquiry said Friday that officials were surprised by Taser's reaction.

"We didn't expect this type of action to be taken," said Chris Freimond. "Mr. Braidwood is an experienced and respected jurist."

The Braidwood Inquiry was sparked by the 2007 death of a Polish immigrant at Vancouver International Airport who stopped breathing after being shocked five times by Royal Canadian Mounted Police officers.

Braidwood was charged by the provincial government with looking into Taser use in the province, where Tasers were introduced in Canada. Braidwood was also asked to provide a complete record of the circumstances surrounding the airport death, which is still ongoing.

The Braidwood Inquiry involved public testimony by Taser executives, police officers and opponents of the stun gun. It examined medical research, testimony from doctors, and test studies supportive and critical of the stun gun.

In its petition, Taser said it was treated unfairly by the inquiry. It accused officials of overlooking key information, including scientific studies and expert testimony, in favor of the stun gun.

Taser points specifically to the lawyer and to the chief overseer of medical and scientific research, saying any conclusions in the report are tainted by their bias. It asks the court for an injunction restraining Braidwood from making any conclusion about the medical safety or risks of the stun gun.

Taser also says the Braidwood Inquiry violated law by releasing the report without first giving Taser the chance to provide a response to the findings.

Avoid aiming Taser at chest: Manufacturer to cops

Vancouver Province: October 8, 2009

In a Sept. 30 training bulletin, TASER International instructed users to — in the interests of evading "controversy" — refrain from targeting the chest area to avoid impact to the heart.

"When possible, avoiding chest shots with electronic control devices avoids the controversy about whether ECDs do or do not affect the human heart," said the bulletin, which is posted on the company's website.

While maintaining the weapons are safe, the company recommends users aim for the abdomen, legs or back. The website also includes a diagram highlighting the recommended target areas on a suspect's body.

Both the Vancouver and Calgary police confirmed Thursday they'll be following the new avoid-the-heart directive.

"We are training to aim at lower centre mass as of yesterday," said Vancouver Police Department spokesman Const. Lindsey Houghton.

Previously, officers were trained to aim at the upper chest and upper back area, he added.

RCMP have not yet announced whether they are implementing the new guidelines.

Calgary police Chief Rick Hanson sent a memo to the rank-and-file Wednesday, announcing the change.

"Clearly there has been more and more medical research that has been conducted that has caused a change in this target," he said.

TASER International said the new directive improves the safety of the weapons and enhances officers' ability to defend themselves against post-event lawsuits.

Should a suspect die of sudden cardiac arrest after being hit by a Taser in the chest area, it would place the officer, the law enforcement agency, and TASER International in the "difficult situation of trying to ascertain what role, if any, the Taser ECD could have played," said the company bulletin.

Calgary police field training co-ordinator Staff Sgt. Chris Butler said Calgary officers will aim for the diaphragm or lower whenever possible — but he stressed that, in an emergency situation, an officer might not be able to hit exactly where he aims.

"Even if the officer's intent is to put the laser dot on the abdomen and fire the Taser, by the time the darts actually impact the subject's body, it could be in a different position," Butler said.

The use of Tasers by police has been highly controversial in Canada, with more than two dozen deaths being associated with the devices.

Zofia Cisowski, the mother of Robert Dziekanski, has filed a lawsuit in B.C. Supreme Court against the federal and provincial governments, the Vancouver International Airport and the four RCMP officers involved in her son's death.

Retired judge Thomas Braidwood, who is overseeing an inquiry into the death, this summer released a series of recommendations, including banning the use of Tasers by police unless a suspect is physically harming someone or about to while committing a criminal act.

Closing submissions continue this week in the inquiry. The Alberta government recently completed testing on all of the Tasers used by municipal forces in the province. In the second round of tests, 970 of the weapons were examined to see if they were meeting manufacturer's specifications. Of those, 88 — about nine per cent — were operating outside of specifications, said Alberta solicitor general spokeswoman Michelle Davio.

In the first round of testing, about 12 per cent of the weapons failed to operate as specified. Those found not to comply have been returned to police departments who will decide to get rid of the weapons or fix them. If they choose to repair the Tasers, they will have to be re-tested before put back into use, Davio said.

Taser issues new warning on shooting into chest area.

Privateofficernews: October 22, 2009

The maker of TASER stun guns is advising police officers to avoid shooting suspects in the chest with the 50,000-volt weapon, saying that it could pose an extremely low risk of an "adverse cardiac event."

The advisory, issued in an Oct. 12 training bulletin, is the first time that TASER International has suggested there is any risk of a cardiac arrest related to the discharge of its stun gun.

But TASER officials said Tuesday that the bulletin does not state that TASERs can cause cardiac arrest. They said the advisory means only that law-enforcement agencies can avoid controversy over the subject if their officers aim at areas other than the chest.

The recommendation could raise questions about whether police officers will find it more difficult to accurately direct the probes emitted by a TASER gun at a recommended body area in order to subdue a suspect. TASER officials say the change won't hinder officers' ability to use TASERs.

In a memo accompanying the bulletin, TASER officials point out that officers can still shoot the guns at a suspect's chest, if needed.

Police departments across the United States and in Canada and Australia reacted immediately to the bulletin, with some ordering officers to follow TASER's instructions and begin aiming at the abdomen, legs or back of a suspect.

Officials with the Phoenix Police Department, one of the first in the country to arm all its officers with TASERs, said Tuesday that the new guidelines are being adopted by trainers who are reviewing departmental policy for possible changes.

Critics, including civil-rights lawyers and human-rights advocates, called the training bulletin an admission by TASER that its guns could cause cardiac arrest. They called it a stunning reversal for the company which for years has maintained that the gun was incapable of inducing a cardiac arrest.

Scottsdale-based TASER insisted that the revision admitted no risk of cardiac arrest and served only as risk-management advice for law enforcement.

In the past, TASER has cautioned that use of its stun gun involves risk inherent in police-suspect conflicts, including the risk that suspects fall after being struck by a TASER.

"TASER has long stood by the fact that our technology is not risk-free and is often used during violent and dangerous confrontations," TASER Vice President Steve Tuttle said in an e-mail. "We have not stated that the TASER causes (cardiac) events in this bulletin, only that the refined target zones avoid any potential controversy on this topic."

TASER's training bulletin states that "the risk of an adverse cardiac event related to a TASER. .. discharge is deemed to be extremely low." However, the bulletin says, it is impossible to predict human reactions when a combination of drug use or underlying cardiac or other medical conditions are involved.

"Should sudden cardiac arrest occur in a scenario involving a TASER discharge to the chest area, it would place the law-enforcement agency, the officer and TASER International in the difficult situation of trying to ascertain what role, if any, the TASER. .. could have played," the bulletin says.

The bulletin recommends that when aiming at the front of a suspect, the best target for officers is the major muscles of the pelvic area or thigh region. "Back shots remain the preferred area when practical," it says.

For years, TASER officials have said in interviews, court cases and government hearings that the stun gun is incapable of inducing ventricular fibrillation, the chaotic heart rhythm characteristic of a heart attack.

The guns are used by more than 12,000 police agencies across the country, including every major law-enforcement agency in the Valley. Many authorities credit the weapon with preventing deaths and injuries to officers and suspects.

Advocacy groups such as Amnesty International allege that TASER guns are often used by police as a compliance tool on unarmed individuals who pose no deadly threat, who are drunk or on drugs and simply quarrel with officers.

Mark Silverstein, legal director of the Colorado American Civil Liberties Union, who has tracked TASER issues for years, said the bulletin means that police departments should now be asking questions about liability and reconsider how the stun gun is used.

"This is further evidence that law-enforcement agencies need to stop and ask if they have been sold a bill of goods," he said. "This (training) bulletin confirms what critics have said for years: that TASER has overstated its safety claims.. .. (It) has to be read as if TASERs can cause cardiac arrest."

Since 2001, there have been more than 400 deaths following police TASER strikes in the United States and 26 in Canada. Medical examiners have ruled that a TASER was a cause, contributing factor or could not be ruled out in more than 30 of those deaths.

The training bulletin is drawing significant attention in Canada, where controversy erupted after the 2007 death of a Polish immigrant at Vancouver International Airport. The man stopped breathing after being shocked five times by Royal Canadian Mounted Police officers.

A Canadian government investigation in July concluded that TASER stun guns can cause death, spurring law-enforcement agencies across the country to put severe new restrictions on how and when police there can use the weapons.

In view of TASER's bulletin, the Mounties revised policies to urge officers to avoid firing at suspects' chests.

124 · CONDUCTIVE ELECTRONIC WEAPONS AND THEIR FAULTS

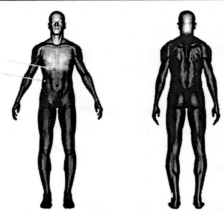

PREFERRED NEW TARGETING LOCATIONS
(Dziekanski's impact area shown for reference)

Police hedge on Taser use after new rules
CBC News: December 7, 2009

Police officers are rethinking their use of stun guns, after Taser International advised them to stop aiming at the chest because of a slight risk of cardiac arrest when the electrified darts hit there.

The company has advised the police to instead aim for the belly or the back of the legs.

Taser International has always assured police officers that the stun gun can never cause problems with the heart, no matter how many times it is used on someone.

Now the company says, just to be safe — from a liability point of view — police should avoid firing at the chest.

"The entire playing field has been altered," said Tony Simioni, president of the Edmonton Police Association.

He said that even before this recent policy, some police officers had begun leaving their Taser in their locker "based on the level of scrutiny, just the overall perception that this Taser may be more trouble than it is worth."

"Now there's an increasing tendency not to use the Taser at all because the deployment instructions are unrealistic, in their view," Simioni said.

Mike Sutherland, president of the Winnipeg Police Association, said the directive now puts his members at risk if they fire the stun gun and it hits the chest.

"I think there's a possibility that we may see an increase — especially given the controversy surrounding the Taser — that our members dragged into court cases where liability will become an issue," Sutherland said.

More Details of Commissioner of Complaints Report on Dziekanski

250 News: December 8, 2009

Vancouver, B.C.- Saying the versions of events given by the RCMP members involved in the Dziekanski case, were not credible, the Commissioner for Complaints against the RCMP has released his final report on the death of Robert Dziekanski. Here are the key findings and recommendations:

"Overall, I found that the conduct of the responding members fell short of that expected of members of the RCMP. The members demonstrated no meaningful attempt to de-escalate the situation, nor did they approach the situation with a measured, coordinated and appropriate response. The failure of the senior member to take control of the scene, communicate with and direct the more junior and inexperienced members negatively manifested itself throughout the interaction with Mr. Dziekanski.

I do not accept the version of events as presented by the four responding RCMP members. The statements provided by the members are sparse in terms of detail of the events and the thought processes of the members as events unfolded. When tracked against the witness video, the recollections of the members fall short of a credible statement of the events as they actually unfolded. The fact that the members met together prior to providing statements causes me to further question their versions of events.

An issue inextricably linked to the incident is the use of a conducted energy weapon (CEW), also known as a TASER®, by an RCMP member during the arrest of Mr. Dziekanski. The CEW is a prohibited firearm pursuant to the regulations under the Criminal Code of Canada.1 Debate pertaining to the overall appropriateness of the use of CEWs by police had been ongoing for some time prior to the YVR incident (and has been previously commented on by the Commission as indicated below), but this particular use of a CEW focused considerable attention and scrutiny on appropriate CEW usage and the nature of the CEW as a weapon.

Overall, while I found that the IHIT investigation was unbiased, I did find a number of issues involved in the IHIT investigative processes. I also found issues with the RCMP's media releases related to this incident. It is essential that the RCMP develop a media and communications strategy specifically for in-custody death investigations that recognizes the need for regular, meaningful and timely updates to the media and to the public. In addition, the media and communications strategy should include a publicly available general investigative outline of the steps to be taken and the anticipated timeline for each step. "

The Commission's Findings and Recommendations:

As a result of my investigation, I made a number of findings and recommendations that I believe will assist the RCMP in enacting/reviewing policies and shape training to ensure that a tragic situation such as this is not repeated.

CPC Final Report Findings:

1.Finding
The RCMP members involved in the arrest of Mr. Dziekanski were in the lawful execution of their respective duties and were acting under appropriate legal authority.

2. Finding

In light of the information possessed by the RCMP members responding, the decision to approach Mr. Dziekanski to deal with the complaints was not unreasonable. At any point a member of the travelling public or an employee at YVR could have happened upon Mr. Dziekanski. As evidenced by the multiple calls to 911, it was incumbent upon the RCMP members to ensure a safe environment for the public and employees using the airport facility and to halt the disturbance being caused by Mr. Dziekanski.

3. Finding

To ensure a coordinated approach to Mr. Dziekanski, Corporal Robinson should have taken control and directed the other responding members to ensure that each was aware of the intended response and to ensure that each communicated with the others as the events unfolded.

4. Finding

Prior to deploying the CEW, Constable Millington should have issued the required warning/challenge to Mr. Dziekanski as required by RCMP policy, notwithstanding the fact that Mr. Dziekanski appeared not to understand the English language.

5. Finding

Because no significant attempts were made by the RCMP members present to communicate with Mr. Dziekanski, to obtain clarification of information pertaining to Mr. Dziekanski's situation, or to communicate among themselves, deployment of the CEW by Constable Millington was premature and was not appropriate in the circumstances.

6. Finding

Constable Millington cycled the CEW multiple times against Mr. Dziekanski when those subsequent cycles were not known by him to be necessary for the control of Mr. Dziekanski.

7. Finding

The multiple cycles of the CEW against Mr. Dziekanski when no significant effort was made to determine the effect of the CEW on

Mr. Dziekanski was an inappropriate use of the CEW.

8. Finding

Corporal Robinson did not adequately monitor Mr. Dziekanski's breathing and heart rate.

9. Finding

Because Corporal Robinson did not know the qualifications of Mr. Enchelmaier, he should not have allowed him to provide first aid or actively monitor Mr. Dziekanski's condition. That task should have been performed by the RCMP members themselves. Corporal Robinson, therefore, failed to provide adequate medical care to Mr. Dziekanski.

10. Finding

The handcuffs should have been removed from Mr. Dziekanski when the members recognized that he was unconscious and in distress and no immediate threat to the members was perceived. At a minimum, they should have been removed immediately upon the initial request of medical personnel.

11. Finding

The failure of Corporal Robinson to take control of the scene, communicate with and direct the more junior and inexperienced members negatively manifested itself throughout the interaction with Mr. Dziekanski.

12. Finding

I do not accept as accurate any of the versions of events as presented by the involved members because I find considerable and significant discrepancies in the detail and accuracy of the recollections of the members when compared against the otherwise uncontroverted video evidence. In their statements, the members indicated in responses to numerous questions that they could not recall the detail of the events as they unfolded. The fact that the members met together and with the SRR prior to providing statements causes me to question further their versions of events.

13. Finding

The conduct of the responding members fell short of that expected of members of the RCMP by the Canadian public and by RCMP policies. The members demonstrated no meaningful attempt to de-escalate the situation, nor did they approach the situation with a measured, coordinated and appropriate response.

14. Finding

The members failed to adequately comply with their training in CAPRA and IM/IM to assess the behaviour of Mr. Dziekanski, and therefore the risk posed by him. As a result, the level of intervention went beyond what was necessary and acceptable, contrary to the RCMP's IM/IM and CAPRA model.

15. Finding

Because the RCMP positions the CEW as an intermediate weapon and trains its members that it is appropriate to use the CEW in response to low levels of threat because it is a relatively less harmful means of controlling a subject, the responding members did not fully appreciate the nature of the CEW as a weapon and it was resorted to too early.

16. Finding

Although IHIT did engage the services of a use of force expert, that expert was not provided with adequate direction in terms of the questions to be considered, the scope of his work or the terms of reference he was to consider.

17. Finding

Corporal Robinson, as an involved member, should not have been allowed to attend the IHIT briefing held at the Richmond Detachment on October 14, 2007. Sergeant Attew failed to ensure that only appropriate RCMP members were present during the briefing.

18. Finding

The responding RCMP members meeting alone at the YVR sub-detachment office following the death of Mr. Dziekanski was inappropriate.

19. Finding
An SRR should not have been permitted to meet alone with Constable Millington prior to the IHIT investigator.

20. Finding
If for no other reason than to be fair to the responding members and give them an opportunity to address the significant and readily apparent discrepancies between their versions of events and the video, it would have been appropriate to provide the responding members with an opportunity to view the Pritchard video prior to taking further statements from them.

21. Finding
The responding members did not keep adequate notes of the incident involving Mr. Dziekanski.

22. Finding
No bias or partiality toward the involved RCMP members was present in the IHIT investigation of the death of Mr. Dziekanski.

23. Finding
The RCMP should have released certain information to the media which would have served to clarify information pertaining to the death of Mr. Dziekanski and correct erroneous information previously provided without compromising the IHIT investigation.

CPC Final Report Recommendations:

1. Recommendation
The RCMP should review the CEW quality assessment program as currently in effect and consider whether it should be enhanced to ensure that a high degree of confidence may be placed in the performance of in-service CEWs.

2. Recommendation
The RCMP should continue to be involved in and stay abreast of current independent research on the use and effects of the CEW.

3. Recommendation

Notwithstanding the fact that the RCMP has (as of January 2009) amended its policy such that the use of the CEW is to be used in response to a threat to officer or public safety as determined by a member's assessment of the totality of the circumstances being encountered, the RCMP should clarify for its members and the public what the appropriate circumstances for using the CEW are and what threat threshold will be utilized to assess the appropriateness of such use.

4. Recommendation

The RCMP should consider a review of its training to ensure that its members are well versed in the potentially dangerous nature of the weapon and to ensure that training provided to members assists them in appropriately assessing the circumstances in which deployment of the CEW is justified, bearing in mind the degree of pain inflicted on the subject during the CEW deployment and the potential outcome of such deployment.

5. Recommendation

The RCMP should consider designing and implementing training for its members in techniques to communicate with persons who cannot meaningfully communicate with them.

6. Recommendation

The RCMP should:

1. Amend its Conducted Energy Weapon (CEW) Usage Reporting Form (Form 3996), to require that information concerning a spark test be captured as part of the CEW usage reporting process (or include such requirement in the forthcoming Subject Behaviour/Officer Response data base); and

2. Edit its Operational policy to emphasize the importance of the spark test and clearly indicate that the spark test is mandatory to confirm proper functioning of the CEW.

7. Recommendation

RCMP detachment familiarization procedures should include a detailed review of available medical facilities and equipment.

8. Recommendation

The RCMP should review its processes and criteria with respect to the initiation of an internal investigation into the conduct of its members to ensure consistency of application across the country.

9. Recommendation

I reiterate my recommendation from my report on the Police Investigating Police (August 2009) that all RCMP member investigations involving death, serious injury or sexual assault should be referred to an external police force or provincial criminal investigation body for investigation. There should be no RCMP involvement in the investigation. If, however, the RCMP continues to investigate such matters, then I recommend that the RCMP implement clear policy directives that all investigations in which death or serious bodily injury are involved and which involve RCMP members investigating other police officers will be considered criminal in nature until demonstrated not to be.

10. Recommendation

If the protocol of SRR attendance is to continue, the RCMP should formalize the role of the SRR to provide clear policy and guidance to ensure that the SRR knows the bounds of his or her involvement and the required protocols with respect to such attendance, and that in all such cases the SRR not meet alone with a subject member in advance of being interviewed by an investigator.

11. Recommendation

I reiterate my recommendation in the Ian Bush decision (November 2007) that [t]he RCMP develop a policy that dictates the requirement, timeliness and use of the duty to account that members are obliged to provide.

12. Recommendation

The RCMP should review its operational policies and procedures to ensure that, particularly in serious cases in which members investigate the actions of other members, processes are available to enable investigator awareness of the nature and depth of detail required during interviews.

13. Recommendation

The RCMP should take steps to ensure that members are aware of the importance of note taking, and that supervisors should be encouraged to regularly review the notes taken by their subordinates to ensure the quality of such documentation.

14. Recommendation

Given the proliferation of recording devices, it is anticipated that incidents in which RCMP members will seek to obtain private video or audio recordings will potentially occur more frequently in the future. Whether the police seize a video or audio recording of an event or obtain it on consent from a member of the public, the police must know and advise the public of the authority under which the video or audio recording is obtained. I recommend that the RCMP provide clarification for members with respect to obtaining video or audio recordings of an event.

15. Recommendation

I reiterate my recommendation in the Ian Bush decision that [t]he RCMP develop a media and communications strategy specifically for police-involved shooting investigations that recognizes the need for regular, meaningful and timely updates to the media and to the public. In addition, the media and communications strategy should include a publicly available general investigative outline of the steps to be taken and the anticipated timeline for each step. I also expand my recommendation to cover all in-custody death investigations.

16. Recommendation

The RCMP should immediately conduct a review of its policies and training regimen to ensure that members are adequately trained with respect to recognizing the risks inherent in, and signs of, positional asphyxia and in taking steps to mitigate those risks.

Mounties too quick to Taser Dziekanski: Report

Vancouver Sun: December 8, 2009

Four RCMP officers who confronted Polish immigrant Robert Dziekanski at the Vancouver airport in 2007 were too quick to deploy a Taser and made no attempt to resolve the situation without violence, a new report says.

Saying the officers fell short of what's expected of Canada's law enforcement team, Paul Kennedy, chairman of the Commission for Public Complaints Against the RCMP, found the use of the Taser against Dziekanski was "premature and inappropriate" and the RCMP members should have provided first aid and monitored Dziekanski's condition.

The report found the members did not attempt to de-escalate the situation nor did they approach the situation "with a measured, co-ordinated and appropriate response."

"The version of events given to investigators by the four RCMP officers involved in the Vancouver International in-custody death of Robert Dziekanski is not deemed credible by the CPC," Kennedy said in a statement.

Kennedy released his report Tuesday into the death at the Vancouver International Airport on Oct. 14, 2007.

The public complaints commission started an investigation in November 2007 to investigate the appropriateness of the response by the RCMP to the complaints concerning Dziekanski's behaviour and the police investigation into Dziekanski's death.

Kennedy has offered a number of recommendations, including having the RCMP review its Taser quality assessment program, training its members in techniques to deal with people who " cannot meaningfully communicate with them."

RCMP Commissioner William Elliott said in a letter to Kennedy the police are waiting for the final report of Thomas Braidwood's inquiry into Dziekanski's death before commenting on the CPC report.

"It has not been the practice for interim reports to be made public and we do not believe it is appropriate for you to do so in this case or more broadly," he wrote.

He said the RCMP has already taken action on many of Kennedy's recommendations.

Kennedy, whose contract ends this month, said he is not impressed with the RCMP's decision to wait for the Braidwood report before responding to his findings.

"There is no excuse for delay," he said. "I can think of no more top of mind issue for the RCMP than this inquiry."

Kennedy said he's disappointed the RCMP didn't start a disciplinary investigation into the situation, noting there's a one-year limitation period to do so. As a result, the officers may not now face any disciplinary action.

"What I've said all along is the RCMP is too slow. The investigation should be done within a year," he said.

Dziekanski, 40, came to Canada to live with his mother. He had been travelling for more than 20 hours and had spent about 11 hours in the international arrivals area of the Vancouver International Airport.

Dziekanski spoke no English and was not provided a translator by customs. He was in a secure area that the public and his mother could not access.

Dziekanski began throwing around his luggage, airport furniture and a computer on a counter.

Seconds after the four Mounties arrived, a Taser was deployed five times over a 31-second period.

Dziekanski died at the scene after being restrained and handcuffed.

The incident was captured on videotape and was the subject of public inquiries. Dziekanski's mother also filed a lawsuit over the RCMP's handling of the case.

In August, Kennedy's watchdog agency called for a halt to the practice of the federal police force investigating its own members in cases of serious injury or death.

Kennedy stopped short of recommending totally independent investigations with no police involvement. Instead, the agency proposed a middle-ground approach to ensuring the integrity of investigations into potential criminal conduct by members of the RCMP.

Citing the need to appear impartial and above suspicion, Kennedy called for an enhanced civilian involvement in the investigations.

Taser timeline in Canada
Canwest News Services: April 1, 2010

Robert Dziekanski's death in 2007 prompted an international outcry, and raised questions about the use of Tasers by Canadian law-enforcement officers.

These are some of the changes to Taser policy that have been made across the country since his death.

June 2008
The Commission for Public Complaints Against the RCMP recommends tighter controls on the use of Taser.

A Parliamentary committee recommends that the RCMP restrict the circumstances in which Tasers are used, that the high-voltage devices be used less and that multiple firings also be restricted.

July 2008

The Saskatchewan Police Commission reverses an earlier decision and no longer supports the idea that all police be equipped with a Taser.

Nova Scotia acts on recommendations from a ministerial review and restricts the circumstances in which Tasers can be used to cases of aggressive resistance or threats to a police officer.

February 2009

The Canadian Police Association and Canadian Association of Chiefs of Police both recommend that all police officers be issued Tasers.

The RCMP tighten the guidelines on Taser use, now requiring that an immediate threat to an officer or the public exist before the weapon is deployed. Changes to training focus on multiple use of the device and the impact on acutely agitated individuals.

July 2009

Alberta enacts new guidelines that require ongoing testing of the devices, and a use-of-force reporting system.

July 2009

British Columbia issues new guidelines in response to the Braidwood inquiry's Phase 1 report. Police should only use the Taser when bodily harm is threatened, when lesser force has proved ineffective or during the enforcement of a federal criminal law.

October 2009

Taser International sends a directive recommending the Taser not be aimed directly at the chest.

March 2010

Ontario announces new guidelines for all police in the province beginning in a few months. Police should refrain from Tasering the elderly and should not aim for the head, throat or genitals, the guidelines say.

First test of the new TASER® X3™
Popular Science: July 15, 2009

The first test of a new product is understandably an exciting time for everyone. No one was more happy than the author to see the first demonstration test of their new X3™. What's even better is the new precautions that are being taken in this video. As you can see in the pictures, the three volunteers with help on either side of them, that's standard. The part that really tickles the author is the person in charge of the test has had the volunteers take off their socks and shoes! The test is being done with bare feet on a non triboelectric surface. The reason for this is shoes create an insulation barrier, and standing in bare feet allows any charges to freely dissipate without affecting the devices output. (1) See pages 27 - 31 for the detailed description.

It may be hard to see the feet with the current print resolution, but the video shows it clearly! In later videos, the volunteers keep their shoes on.

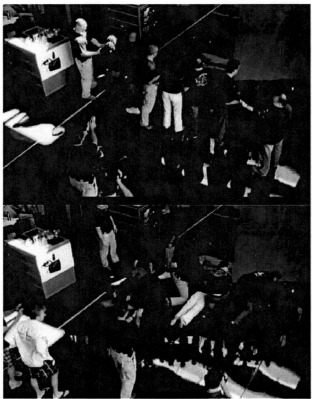

Chapter 9 Fallout of the Inquiry

Mountie who Tasered Robert Dziekanski files libel lawsuit against CBC

Vancouver Sun: November 3, 2009

The Mountie who repeatedly Tasered Robert Dziekanski at Vancouver's airport two years ago has filed a libel lawsuit against the CBC.

The legal action by Constable Kwesi Millington claims that CBC's coverage of the incident since Nov. 4, 2007 has caused him to suffer "serious embarrassment and distress" and has caused him "public ridicule."

Millington claims CBC's coverage has "seriously injured" his reputation, causing him to seek punitive and aggravated damages against the CBC.

The former Richmond Mountie deployed his Taser five times – the last two in "push stun mode" – during the incident on Oct. 14, 2007.

Four officers arrived at the scene to investigate complaints of a possibly drunk man throwing around furniture. Seconds after arriving, the officers confronted Dziekanski, gave him conflicting commands, then Tasered him.

Dziekanski died shortly after he was restrained and handcuffed by the four officers, who testified at the Braidwood inquiry that they feared for their safety after Dziekanski picked up a stapler from a nearby desk.

The four officers were reassigned after the incident – Millington was transferred to Toronto.

The senior officer who attended the call, Cpl. Benjamin (Monty) Robinson, had been reassigned to the 2010 Integrated Security Unit but was suspended with pay after motorcyclist Orion Hutchinson, 21, was fatally struck by a Jeep in Delta last year.

Police recommended that charges of impaired driving and impaired driving causing death be laid against Robinson but the matter still is being reviewed by the Crown.

Thomas Braidwood, the former judge who headed the Braidwood inquiry, currently is preparing his final report on the circumstances that led to Dziekanski's death.

RCMP tarnished by Dziekanski case: expert
CBC News: April 2, 2010

It's going to take more than money and an apology to redeem the RCMP's image following the Robert Dziekanski case, some critics say.

The announcement of a financial settlement and a formal apology from the RCMP was welcomed Thursday by Sofia Cisowski, the mother of Dziekanski, the Polish immigrant who died at Vancouver airport in 2007 after being confronted by four officers and stunned several times with a Taser.

"I really seriously doubt the majority ... are convinced that this in any way compensates for what happened in the Dziekanski matter," Rob Gordon, a criminologist at Simon Fraser University, told CBC News.

The officers repeatedly used a Taser to subdue 40-year-old Dziekanski and then pinned him down for several minutes just before he died. A months-long judicial inquiry headed by Justice Robert Braidwood focused on questions of excessive use of force in dealing with Dziekanski. Braidwood's findings have not yet been made public.

"If they had [apologized] within weeks of the event — and I think there was ample opportunity for them to do that — it would have carried much more weight," said Gordon.

"Instead, they've dragged the Dziekanski family and indeed the province through 2½ years of investigation and inquiry before they finally fess up."

RCMP apology falls short
Edmonton Journal: April 2, 2010

RCMP Deputy Commissioner Gary Bass offered an apology Thursday to Zofia Cisowski, the mother of Robert Dziekanski.

"On behalf of the RCMP, I want to apologize for our role in the tragic death of your son. Your son arrived from Poland eager to begin a new life here in Canada. We are deeply sorry he did not have that opportunity."

Considering the ramifications of this dark chapter in contemporary Canadian history, it should have been Commissioner William Elliot who made that statement. You have to wonder what pressing matters impelled the leader of the force to download public responsibility to an underling for one of the blackest marks ever chalked up against our national police force. But then, Canadians have become used to being disappointed in an institution that has suffered numerous setbacks in recent years at its own hands.

The apology should have been extended to all Canadians, especially those who have doggedly fought to reveal the truth. Paul Pritchard, who digitally captured Dziekanski and the police on his video camera deserves a special vote of thanks for sterling and indispensable citizen involvement.

On a recent visit to Edmonton, Elliot opined that the Mounties' "transformation strategy" is working. Let's hope whatever that might actually mean includes never again repeating the sorry events of the Dziekanski affair, which has left a man dead and a nation embarrassed internationally. Throughout the 2½ years since the 40-year-old immigrant lost his life after being Tasered five times by RCMP officers, those who have questioned the force have been systematically tarred as little short of unpatriotic. Now we know who was right and who was wrong. In addition to the RCMP mea culpa, letters of apology were also tendered by the B.C. solicitor general and the Canada Border Services Agency, organizations that also bear a degree of responsibility for their respective actions and stonewalling activities.

In tandem with the apology was a civil financial settlement tendered to Ms. Cisowski by the federal government -- which means taxpayers have paid materially for this nonsense. Bass said Thursday that he hopes the apology and money "marks the beginning of the healing process for Mrs. Cisowski, the RCMP and the public. It is critically important that the public has confidence in the police if they are to be able to work collaboratively to ensure public safety. We hope that the steps announced today will work toward this goal."

That's a wish that all of us might share. For her part, Cisowski was nothing short of gracious in her appearance at the same Richmond, B.C., news conference. "There was not a single day I did not cry and analyze what could have been done to avoid this tragedy," she said before losing it to her emotions. The RCMP will contribute $20,000 for a scholarship in Dziekanski's name at Thompson Rivers University in Kamloops -- "my son's legacy," she calls it, adding it will be part of her healing process.

Canadians can take some comfort in affirming that, in the end, the system delivered a measure of justice, if over a long, rocky and less than edifying process. Lives cannot be returned, however, and the true measure of the awful Dziekanski affair will be whether lessons -- from Taser use to personal accountability and professionalism -- will be learned. For many of us, the horrific 11 hours spent by Robert Dziekanski in the Vancouver airport will remain an indelible stain on the good name of Canada for years to come. A man and his family were badly let down, and we won't forget easily, nor should we.

Braidwood's final report in gov't hands

The Province: May 21, 2010

Thomas Braidwood's long-awaited final report on the Taser-related death of Robert Dziekanski at Vancouver airport has been delivered to government.

But the public will have to wait at least a month to read the report from Braidwood's inquiry -- and may never get to see parts that could be excised for "privacy" reasons.

B.C. Attorney-General Michael de Jong accepted the report Thursday from ex-Supreme Court justice Braidwood, whose findings for the inquiry's second phase focus on the October 2007 death of Dziekanski at the airport.

Dziekanski's mother, Zofia Cisowski, attended much of the second phase of the hearings, although she left during viewings of a video shot by Paul Pritchard of her son's death in RCMP custody.

Braidwood's goal in the second phase was to "provide Dziekanski's family and the public with a complete record of the circumstances of his death."

Cisowski recently accepted an apology by B.C. RCMP brass, and an undisclosed financial settlement, in return for dropping her planned civil suit against the RCMP and the federal government. However, the B.C. Court of Appeal earlier ruled that Braidwood is free to make findings of misconduct against the four Mounties involved in Dziekanski's death.

The first phase of Braidwood's inquiry into the use of " conducted-energy weapons," or Tasers, in B.C. was released in July, 2009. Dave Townsend, spokesman for the attorney-general, said the second report likely won't be released until a year after the first report, in mid-July 2010.

After cabinet views the document, it must be vetted by B.C.'s privacy commissioner.

Braidwood inquiry finds officers' use of Taser not justified; AG appoints special prosecutor

Vancouver Sun and Canwest News Services: June 18, 2010

A special prosecutor has been appointed to review the actions of the four RCMP officers involved in the Tasering of Polish immigrant Robert Dziekanski.

B.C. Attorney-Solicitor General Michael de Jong made the announcement Friday following the release of the final report by the Braidwood Inquiry, which reviewed the 2007 Tasering and death of Dziekanski.

In his report, the head of the inquiry, Commissioner Thomas Braidwood, said what he called the "shameful conduct" of the officers was not justified.

Braidwood, a retired appeal court judge, bluntly condemned the actions of the four RCMP officers who, responding to a report of a violent drunk at the Vancouver International Airport in October 2007, repeatedly shocked Dziekanski with a Taser as he writhed in pain.

Braidwood, releasing his final report on the high-profile case Friday, quoted Dziekanski's final words before his death, spoken in Polish: "Leave me alone. Did you become stupid? Have you gone insane? Why?"

"In my view, Const. (Kwesi) Millington was not justified in deploying the weapon against Mr. Dziekanski, given the totality of the circumstances he was facing at the time," Braidwood said.

"Similarly, Cpl. (Benjamin) Robinson was not justified in instructing him to deploy the weapon."

Braidwood dismissed as false the RCMP officers' claims that they were forced to fight with Dziekanski.

"The initial claims by all four officers that they wrestled Mr. Dziekanski to the ground were untrue," said the report.

"In my view they were deliberate misrepresentations, made for the purpose of justifying their actions."

Dziekanski, who spoke no English and had never been on a plane before, was unable to find his mother upon arriving at the airport. He remained in a secure customs area for nearly 11 hours and then, appearing dazed and delirious, began throwing around furniture, prompting the 911 call.

Moments after four RCMP officers — Millington, Robinson, Const. Bill Bentley and Const. Gerry Rundel — arrived on the scene, Dziekanski was jolted five times with a Taser.

He died minutes after he was restrained and handcuffed face-down on the airport floor.

The incident, captured on amateur video, resulted in an international public outcry.

The officers testified at the inquiry that they believed Dziekanski intended to attack.

But Braidwood condemned their testimony, saying, "I do not believe that either of these officers honestly perceived that Mr. Dziekanski was intending to attack them or the other officers.

"They approached the incident as though responding to a barroom brawl and failed to shift gears when they realized that they were dealing with an obviously distraught traveller," he wrote in the report. "Neither officer carried out an appropriate reassessment of risk immediately before deployment of the weapon."

He found Dziekanski was compliant with police commands and did not brandish a weapon, despite grabbing a stapler.

"Mr. Dziekanski did not bring this on himself," Braidwood said.

While Braidwood said that the officers should have tried to calm down the situation without resorting to violence, he stopped short of saying the RCMP officers' actions amounted to misconduct.

"I think I was blunt enough, full enough, and hopefully accurate enough that those reading it can draw their own conclusions," he told reporters.

"This tragic case is, at its heart, the story of shameful conduct of a few officers. It ought not to reflect unfairly on the many thousands of RCMP and other police officers who have, through years of public service, protected our communities and earned a well-deserved reputation in doing so."

Braidwood also levelled harsh criticisms against the RCMP media relations officers responding to the case.

"I conclude that some of the RCMP's reports were factually inaccurate but not intentionally misleading," he said.

"It is not in dispute that some of the RCMP's public disclosures about the Dziekanski incident, during the early stages of the criminal investigation, were factually inaccurate," he wrote.

"When the RCMP became aware of these inaccuracies, the officer in charge of (the Integrated Homicide Investigation Team) decided not to correct them."

Braidwood also said there is an inherent "conflict of interest" when the RCMP investigates a Mountie-related death.

"Many members of the public are understandably suspicious of such investigations, no matter how thorough and impartial they turn out to be," he said, recommending the establishment of a new civilian unit to investigate such incidents.

"If the criminal investigation had been conducted by a body at arm's-length from the RCMP, that body would have been, and would have been perceived by the public to have been, impartial," he said.

To that end, De Jong also said Friday that, within the next year, the B.C. government will create a new civilian-led unit to investigate all independent municipal police- and RCMP-related deaths and serious incidents across B.C.

"Mr. Braidwood has done a tremendous job of unravelling and probing all the circumstances surrounding the tragic death of Robert Dziekanski," said de Jong. "B.C. agrees with the intent, principal and purpose of each of the report's recommendations."

De Jong said the province intends to set up an Independent Investigations Office (IIO) within a year.

"Moving to an IIO model will help prevent in future what played out during the inquiry and is highlighted in the commission's report - a number of discrepancies between what RCMP officers told investigators in 2008 and what came out at the inquiry," de Jong said.

"Since receipt of this report, the assistant deputy attorney general, Bob Gillen, has determined, and I agree, that it is appropriate to appoint a special prosecutor to review this matter."

The Criminal Justice Branch of the Ministry of Attorney General later Friday named Richard C.C. Peck as the independent special prosecutor.

Shortly after the release of Braidwood's report, RCMP Commissioner William Elliot admitted the force's handling of the fatal incident "failed at many levels" and the events should have unfolded differently.

"It is clear our policies and training in place at the time were deficient," he told reporters.

"We acknowledge that the actions of our members who dealt with Mr. Dziekanski also fell short."

Elliot began the news conference by offering Dziekanski's mother "our sincere condolences on the death of her son Robert and to apologize unconditionally for the role the RCMP, including individual members of the RCMP, played in his tragic death."

Elliot admitted, responding to a reporter's question, that the RCMP "messed up," made mistakes and didn't apologize soon enough.

He said the incident undermined the public confidence in the force.

"Public confidence is the cornerstone of effective policing," Elliot said. "Canadians will not support us when they don't trust us."

He said the force has already made changes in training, policy and procedures involving the use of force.

He also said the RCMP welcomed the B.C. government's plan to establish an civilian agency to investigate police-related deaths and other serious incidents.

And the force will cooperate with a special prosecutor appointed to review the original charge approval decision for the four Mounties involved in Dziekanski's death.

He said one of the officers was suspended for an unrelated incident and the other three are assigned to administrative duties.

The Braidwood Inquiry's first report, released last July, examined the use of Tasers — known as conducted-energy weapons — by police, sheriffs and corrections staff in British Columbia.

During the inquiry, a number of medical experts expressed concerns about the risks associated with Taser use, while police officers testified the use of the devices saves lives.

Since Dziekanski's death, the RCMP has made a number of changes to its Taser policies, including restricting the weapon's use to incidents threatening officer or public safety, annual recertification for trained users and enhanced reporting on all use-of-force incidents by RCMP officers.

In April, the RCMP apologized to Dziekanski's mother, Zofia Cisowski, telling her they were "deeply sorry" for their role in her son's death.

"On behalf of the RCMP, I want to apologize for our role in the tragic death of your son," said RCMP Deputy Commissioner Gary Bass at the time. He stopped short of acknowledging the RCMP was to blame.

Along with the apology, Cisowski was also given an undisclosed financial settlement.

For her part, Cisowski dropped a lawsuit for damages against the RCMP, the four officers involved, the Canada Border Services Agency and the Vancouver Airport Authority.

On Friday, Dziekanski's mother said she was happy with the attorney general's announcement.

"I'm happy and tired," she told reporters. "I'm just so excited. It's a difficult day for me today."

De Jong thanked and praised Zofia during his news conference.

"You are a brave lady and I think British Columbians and Canadians have seen this first hand. I thank you," he said.

"The human tragedy of this was staggering - that someone could get lost in an airport."

He added: "'Why' was the question Mr. Dziekanski asked and 'why' was what Canadians asked."

De Jong said he has already contacted his federal counterpart, expressing the B.C. government's support to implement the changes recommended by Braidwood.

NDP Vancouver Kingsway MP and public safety critic Don Davies called Braidwood's report "excellent," adding it struck the proper balance in being critical but fair.

Liberal critic and former B.C. attorney general Ujjal Dosanjh said Braidwood's report, along with the Major report on Air India, released a day earlier, indicate the RCMP has some serious systemic problems.

"The RCMP is in need of repair," he said.

Dosanjh said he believes needs to be accountable to a civilian board and the B.C. Police Complaints Commission so it can be held accountable.

He pointed out that one-third of the RCMP in Canada are employed under contracts in B.C. but are not accountable to the B.C. government.

Taser International calls for Braidwood report to be overturned

Canadan Press: July 5, 2010

A report by the head of a B.C. public inquiry should be overturned because it unfairly concluded Tasers can be deadly without first giving the weapon's manufacturer a chance to refute those findings, says a lawyer for Taser International.

The Arizona-based company is challenging the findings of a report released by commissioner Thomas Braidwood last July - the first of two reports stemming from the death of Robert Dziekanski - that concluded Tasers can kill and recommended restrictions on their use.

Taser's lawyer, David Neave, told B.C. Supreme Court on Monday that Braidwood owed the company a chance to review the report before it was released to the public so it could respond to those findings, especially given their potential impact to Taser's reputation. Neave also insisted Braidwood heard nothing during several weeks of hearings in 2008 to support the conclusion that Tasers can cause serious injury or death.

"On one side of the fairness coin, there's a duty from the commission to Taser," Neave said. "The other side of that coin is a legal right to procedural fairness to Taser."

"The findings that a Taser can cause death are unreasonable. The findings are not supported by the evidence that was known to the commissioner."

In fact, Neave argued that Braidwood heard a significant amount of evidence that showed the opposite - that there is no evidence Tasers can adversely affect the human heart.

Neave complained that Braidwood's report only mentions 57 of the 174 reports that Taser provided to the commission.

The report followed the first phase of a public inquiry after Dziekanski's death at Vancouver's airport in October 2007, when he was confronted by four RCMP officers and repeatedly stunned with a Taser.

While Braidwood said there wasn't enough medical evidence on the precise risk of an electrical shock from a Taser, he said there was sufficient evidence to conclude the weapons have the capacity to affect the heart and cause a fatal arrhythmia.

He also questioned some of the studies and statistics provided by Taser, saying their methodology and results were insufficient to support the company's claims that the weapons can't affect the heart.

Neave told court on Monday that Braidwood's report has hurt the company's business around the world. He referred to an affidavit from co-founder Thomas Smith, who said the report caused customers to question the safety of Tasers and scuttled a multimillion-dollar contract in Africa earlier this year.

"Clearly there has been a negative impact from the decisions in the report," Neave said. "There has been an economic backlash."

Earlier this year, the provincial government asked the court to throw out Taser's petition, arguing the company had no right to challenge Braidwood's report.

Judge Robert Sewell allowed the case to proceed, but forced Taser to remove allegations of bias against the commission's lead lawyer and its medical adviser.

The province's lawyers are expected to repeat their arguments that recommendations from a commission such as Braidwood's aren't subject to judicial review and that the province had no obligation to share the report with Taser before releasing it.

They'll also claim Braidwood did, in fact, read all of Taser's material, even if he didn't include all of the company's studies in his 550-page report.

Braidwood was asked about the Taser's legal challenge last month as he released his second report, which dealt with Dziekanski's death in detail.

"I certainly read all their material, I read everything," he said. "It doesn't mean to say I'm going to put it all in a report. Let me just say I disagree with them."

Taser has a history of successfully defending its weapons in court against any claims they can cause or contribute to someone's death.

For example, when an Ohio medical examiner ruled that three men's deaths were in part caused by the effects of Tasers, the company convinced a judge that any suggestion the Taser was to blame should be removed from the autopsy findings.

Taser claims it has lost business after inquiry ignored its submissions
The Province: July 5, 2010

Taser International launched a five-day hearing in B.C. Supreme Court Monday in a bid to get sections of Thomas Braidwood's report into the death of Robert Dziekanski thrown out.

Taser lawyer David Neave told the court that the sections of the report where Braidwood concluded "conducted energy weapons do have the capacity to cause serious injury or death" should be quashed.

Neave said Taser had the right to "procedural fairness" and that Braidwood ignored 117 documents submitted by Taser, and that Taser wasn't given notice of the findings or given the chance to respond.

He said Braidwood's findings were "unreasonable" based on Taser's submissions and were not supported by the evidence presented to the commission.

Dziekanski, a Polish immigrant, died in 2007 at Vancouver International Airport after being Tasered five times by RCMP.

Neave said that Braidwood's findings had a negative impact on Taser's "commercial interests" and had lost a contract worth "tens of millions of dollars" in Africa.

"There has been an economic backlash," he said.

Taser maker challenges Braidwood report
CBC News: July 5, 2010

Lawyers for Taser International will be in B.C. Supreme Court on Monday to challenge the conclusions of the first Braidwood Taser inquiry, including the finding that the stun gun's electric shocks can be fatal.

Commissioner Thomas Braidwood oversaw the two-part inquiry after the death of Robert Dziekanski, a Polish immigrant who died after receiving several shocks from an RCMP Taser at Vancouver's airport in 2007.

In his first report issued in July 2009, Braidwood concluded Tasers could increase the risk of fatal heart failure under certain circumstances, and their use should be restricted.

In a court petition, Taser International responded that Braidwood didn't take into account the information the company provided and that it wasn't given a chance to respond at the inquiry.

The Arizona-based company claims it has been forced to travel the world defending its products since the report was released and has lost a multimillion-dollar contract in Africa.

Suggested stun guns not be used for bylaws

Braidwood made 19 recommendations in his first report to the B.C. government, including that "conducted energy weapons," or stun guns, should not be used to enforce municipal bylaws or provincial laws — but only federal criminal offences.

He also recommended that the current threshold of "active resistance to police" is too low for use of stun guns and a higher threshold should be adopted.

A police officer must believe, he said, that the subject is causing or is about to cause bodily harm and that no lesser-force option or de-escalation technique would be effective before deploying a stun gun.

He also recommended that stun guns only be used in single five-second bursts in most cases — rather than multiple bursts — citing increased medical risks associated with repeated shocks, and that paramedic assistance be requested in every medically high-risk situation.

Braidwood's second report, which focused on Dzeikanski's death, concluded the RCMP were not justified in using a Taser against him and that the officers later deliberately misrepresented their actions to investigators.

Braidwood inquiry Taser findings upheld by court
CBC News: August 10, 2010

The B.C. Supreme Court has upheld findings by the Braidwood inquiry that stun guns can kill.

Taser International had challenged the findings, calling them unfair, unfounded and a black mark on the company's reputation.

But in a ruling released Tuesday morning, B.C. Supreme Court Justice Robert Sewell said the company's arguments hold "no merit."

Sewell said it's clear to him that the inquiry's commissioner, retired justice Thomas Braidwood, had carefully looked at the opinions of medical experts and his findings were reasonable.

The Braidwood inquiry was launched in the wake of the death of Robert Dziekanski, a Polish immigrant, who died at Vancouver International Airport in October 2007 after being shocked multiple times with an RCMP Taser.

The report released last year after the first phase of the inquiry found that the Taser weapons pose a risk of serious injury or death.

It set out recommendations for their use and cautioned against multiple stuns. Both the B.C. solicitor general and the RCMP endorsed the recommendations.

Taser International had asked the court to quash many of the 19 recommendations made in the report.

Taser International alleged the inquiry neglected to enter scientific and medical evidence brought forward by the company.

Taser lawyers argued in court that the company did not have a chance to see the report before it was released and that the conclusions, based on incomplete information, were not supported by the facts.

Taser International has a history of legal action in defending its products and last year said that it had won its 100th dismissal of a liability lawsuit.

Taser's challenge to overturn Braidwood Inquiry dismissed

News.Yahoo.com: August 11, 2010

A B.C. Supreme Court judge on Tuesday tossed out a legal challenge by Taser International, finding no merit in its bid to overturn a conclusion in the report by the Braidwood Inquiry that found stun guns increase the risk of fatal heart failure.

"Obviously, I am very pleased," said retired justice Thomas Braidwood, who presided over the inquiry into the October 2007 death of Robert Dziekanski at the Vancouver airport, shortly after the decision was released. "The court found there was fairness in the report for Taser International. They had full opportunity to present evidence."

Justice Robert Sewell found no merit in the Arizona-based company's argument that the report was unreasonable in concluding that conducted energy weapons can cause death.

"It is quite clear to me that there were presentations made to the commissioner by medical experts and others to the effect that such weapons can cause serious harm and even death in exceptional circumstances," said Sewell, in the decision.

He also found there was little evidence the report had any serious impact on Taser International's business.

"In my view, there is nothing in the Study Commission Report which a fair-minded person would construe as an attack or criticism of the petitioner's reputation as a corporate citizen or its right to carry on business and market its products," he said.

Sewell noted that Braidwood, in his report, had recommended the continued use of conducted energy weapons and mentioned the advantages of using the weapons in situations where police would be required to use more force.

The judge dismissed the petition without costs.

Taser International filed its legal challenge in August 2009 after the release of Braidwood's report following the first phase of the public inquiry. The company contends that a significant body of evidence delivered to the commission to support the safety of the conducted energy weapon was ignored by the inquiry.

Taser lawyer David Neave had argued that company officials submitted 174 academic and medical articles but Braidwood only referred to 60 in his report. He also said that the conclusions and recommendations in the report would put law enforcement and Canadian citizens at risk. Neave said on Tuesday that he was "carefully reviewing" the decision, but had not yet decided on a response.

The second phase of Braidwood's inquiry focused on the circumstances of the death of Dziekanski, a Polish immigrant who was Tasered five times by RCMP officers at Vancouver International Airport. Braidwood released his second report in June, saying that the Taser hits contributed to his death. Four Mounties responded to a 911 call reporting a possibly drunk man throwing around furniture at the airport. Within 30 seconds of police arriving at the scene, Dziekanski had been Tasered five times and handcuffed. He died face-down on the airport floor.

TASER International Wins Judgment and Final Injunction in Patent Infringement Against Stinger Systems

realpennies.com: August 31, 2010

TASER International Inc. announced that on August 30, 2010, the United States District Court for the District of Arizona entered judgment in favor of TASER International Inc. against Stinger Systems Inc. on Count One of TASER's complaint and further ordered that "Stinger is hereby adjudged to have infringed claims 2 and 40 of United States Patent No. 6,999,295 and that "Claims 2 and 40 of the '295 patent are hereby adjudged to be valid and enforceable." The patent infringement claim against Stinger's S-200 is based on TASER's patent on the dual mode "shaped pulse" technology found in the TASER(R) X26(TM), X3(TM), Shockwave, and C2(TM) electronic control device products. The Court's ruling found that the "Flyback Quantum Technology" found in the Stinger S-200 literally infringes TASER's '295 Patent. The Court also entered an order for a Final Injunction that provides that "Stinger, together with its officers, agents, servants, employees, and attorneys, and those persons in active concert or participation with them who receive actual notice of this Final Injunction, are hereby restrained and enjoined, pursuant to 35 U.S.C. Section 283 and Fed. R. Civ. P. 65, from making, using, offering to sell, or selling in or from the United States, or importing into the United States, the S-200 ECDs, either alone or in combination with any other product, and all other products that are only colorably different from the S-200 ECDs in the context of claims 2 or 40 of the '295 patent, whether individually or in combination with other products or as part of another product, and from otherwise infringing, contributing to the infringement of, or inducing others to infringe claims 2 or 40 of the '295 patent." This Final Injunction runs until the expiration of the '295 patent (February 11, 2023) and also applies to a substantial portion of the components for the S-200 ECD.

Florida Gulf Coast University did a report comparing both companies devices, for the National Institute of Justice. "A Qualitative and Quantitative Analysis of Conducted Energy Devices: Taser X26 vs Stinger S200" March 5, 2008.

The CBC / Radio Canada test update
The Canadian Association of Journalists

After Robert Dziekanski's death on October 14, 2007, CBC / Radio Canada decided it was the right time to take a closer look at what kind of testing had been done by Canadian police authorities prior to adopting Tasers.

It took access-to-information requests and six months of repeated calls to the RCMP to learn that little had been done to verify the electrical output of Tasers. Furthermore, in analyzing the research material they used in making decisions, we realized the RCMP had not only relied on some seriously flawed studies, but had accepted all the information from Taser International at face value. This was a strong incentive for the CBC / Radio-Canada to commission its own tests.

In the weeks after Dziekanski's death, editorial writers and citizens in Canada continued to speculate about whether the Taser was to blame. There was no indication that the stun gun used on him was going to be tested.

Taser International maintained that its electric guns would effectively incapacitate suspects and still have a very high margin of safety. As a matter of fact, in 2006, the company proudly showed CBC / Radio Canada every step of the assembly line at its plant in Scottsdale, Arizona.

At that time, a company executive claimed that all Tasers fired the same output, and that none performed outside the specified range. That assurance was repeated when the president of Taser International went to Vancouver to address the Braidwood inquiry examining Dziekanski's death. And once again, a Taser International executive was adamant that police forces need not test their weapons.

This was the cue to test the company's claims.

CBC / Radio-Canada decided to undertake what became the largest independent testing of Tasers ever conducted in the world.

Using freedom-of-information laws and sources, we obtained the studies and tests conducted on Tasers upon which Canadian authorities had relied.

Several experts analyzed the studies and concluded that they were either extremely limited or inaccurate. We also obtained a testing protocol Taser International provided to the Canadian Police Research Centre. This protocol became valuable for establishing our testing procedure and for ensuring that Taser International could not dispute our testing or our results. We then set out to find a lab that would agree to test the weapons.

We hired National Technical Services (NTS), an internationally accredited engineering and testing firm. NTS randomly tested more than 40 units of the most popular model, the Taser X26. They came from several police departments, all being used by U.S. police officers. The results were troubling: more than 10% of the units were either defective or significantly off specifications. Some of them put out 50% more electrical current than the manufacturer's standards, thus significantly reducing the safety margin.

Pierre Savard, a biomedical engineer at the École Polytechnique de Montréal designed the technical procedure for our testing. He analyzed NTS's results (He was not paid for this work, but felt there was a strong public interest). Other electrical engineers then reviewed his work.

As we were working our way through the roadblocks we encountered in getting a test protocol and a lab to do the tests, we were also filing freedom-of-information requests to all the major police forces in Canada to obtain their Taser inventory lists. They included details about units that were returned or replaced by the manufacturer.

We discovered there was a high rate of return: 15 to 30%. This data confirmed some of the allegations from former employees who had testified in a shareholder lawsuit against the company regarding high numbers of defective weapons.

After reading the affidavits from those former Taser employees, we tracked down some of them to verify their statements. In exchange for guarantees of anonymity, they told us there were problems on the assembly line of the X26 in its early years. They said that at times, more than 70% of the units were defective. Still, there was a lot of pressure to deliver them, both from investors and from police forces.

Our test results showed a strong link between the malfunctioning Tasers and their serial numbers. Low serial number units were much more likely to emit too much current.

Most of the problems we witnessed in our own testing were intermittent and couldn't be seen with the naked eye. The malfunctioning Tasers would be virtually impossible to detect by police forces without submitting them to comprehensive testing. And that's one of the big problems. While all electrical appliances -- no matter how innocuous -- must be tested and certified by Canadian Standards Association (CSA) before being sold in Canada, this is not the case for Tasers.

No regulatory body tests or certifies them in Canada, the United States or anywhere else in the world. Even the police forces using Tasers X26's in Canada, like the RCMP and the Vancouver Police, haven't tested them, nor do they test the weapons as they age. They have relied entirely on the manufacturer and on a study conducted on a single unit in Australia. That's right -- one lonely Taser gun. That singular test was considered to be sufficient.

One very important part of the story was the analysis of our test results using an international electrical standard. The analysis conducted by biomedical engineer Pierre Savard, concluded that the risk of death could be very high in certain circumstances. We contacted some members of the international committee responsible for that standard to verify our findings. Right after the broadcast, we found out that Taser was using a 20-year old version of the international standard and was not aware of some important changes included in the 2007 version of that standard.

The Taser testing stories relied heavily on scientists and medical experts.

We contacted eight electrical engineers, several medical doctors and researchers. We also talked to a dozen police officers. We tracked down former Taser International employees and some of the company's distributors. We obtained an interview with the RCMP and the head of the Canadian Police Research Centre. Taser International refused our request for an interview on this story, despite the fact we had interviewed many of the company's executives and officials during the previous two years.

Immediate Reaction and change -- police decide to test their weapons

The morning after the Taser testing stories aired, the Quebec government announced it was ordering all police forces to remove all of the model X26 Tasers manufactured before 2005 and have them tested. They also committed to randomly test newer models.

The RCMP followed with a similar announcement. Within days, most provinces and major police forces across Canada announced they would pull back their early Taser X26's until they were properly tested. Some provinces, like British Columbia, announced there would be regular testing conducted on Tasers from now on.

CBC / Radio Canada aired follow-up news stories about the reactions to the testing. Taser International attacked the credibility and methodology of the study saying it was "not relevant on a medical safety perspective". Ironically, Taser International criticized part of the methodology that came from its own testing protocol. No correction or clarification was published or necessary. Experts who reviewed our test results confirmed the study was valid and significant.

The Taser Test was also part of the series that won the 2008 Michener Award.

Chapter 10 Conclusions & Recommendations

Behavioral analysis, the unanswered questions

Although the Inquiry came out with over 1000 pages of report, it did not answer a key question as to what caused Mr. Dziekanski's bizarre behavior? The clues however can be found in the autopsy report. The Coroner concludes that he had a fatty liver and other signs which is caused by chronic alcoholism. Liver failure has a number of side effects that can explain what Mr. Dziekanski was experiencing. Delirium, confusion and stupor are signs of Acute Liver Failure.

The Toxicology report also confirms liver failure with the lack of glucose and ketone's. As the levels were not detectable, and the point of visible confusion came about 1 hour after he drank the glass of water in the CBSA offices, means his liver failure must have already been in progress. His unknown whereabouts can be explained by him sitting down behind a counter to have a nap, (as he had been awake for 36 hours at this point) or perhaps he was sitting in plain sight and nobody noticed. Perhaps he was thinking his mother would pick him up at the baggage carousel and just wondered around looking for her? Everything is just speculation without video evidence, however after leaving the customs area the symptoms just grew larger. Along with the disappointment that he missed the connection with his mother, those mixed emotions of anger and confusion, would make anyone mad. Add in nicotine and possible alcohol withdrawal, his pale complexion and sweating is a further side effect of sleep deprivation and exhaustion.

Instead of somebody intervening, he was met with four officers. Instead of first trying to diffuse the situation, they had already made up their minds. Three officers were on the scene waiting outside, any one of them could have come in to see what was going on, assess the situation, even try to diffuse the situation. However as the Inquiry found, the officers already decided on the drive in to see the new device in action.

The behavioral analysis of the RCMP officers

What happened afterwords was a mess. The officers themselves must have been in a state of shock or confusion. This said after the fact, as they were now more interested in getting their stories to match than noticing that a person was turning blue and having agonal breathing right in front of them. There were several people during the Inquiry stated that these officers were only following their training. This document is not written to judge the officers, but its obvious that their training was not sufficient on what to do when the new toy does not work as expected?

The behavioral analysis of the RCMP management.

The Braidwood Inquiry found faults with several levels of the RCMP organization. The police forces (perhaps being influenced by the manufacturers guarantee, are also guilty of recklessly demonstrating the devices to the media, stunning any volunteer reporter for the nightly news. However now that something has obviously gone wrong, their willingness to cooperate has to be forced by media pressure.

The behavioral analysis of the manufacturer.

The manufacturers must accept responsibility that they have chosen to proceed with legal action against any and all who questioned whether the product really operates without risk. Yes their a business, and a business's priority is to make money, however their training programs and marketing practise's have over emphasized the risk less safety of the devices. Blatant usage against any volunteer at the Consumer Electronics Show is an example. Taser International's insistence that their devices don't have to be checked absolutely flies in the face of safety when their own ex-employees gave affidavits that there was up to a 70 percent failure rate on certain older models. The newer models have yet to be publicly verified for correct operation. The wording has slowly changed from, no risk, to less risk, to less lethal, what's next?

When the Braidwood inquiry started the author sent the first CEW report to Taser International, and did not get a reply. The author continued to send in updates when information was also submitted to the inquiry. All this was met with deafening silence. The author believes that Taser International has apparently used the information in their X3™ design to address several specific issues raised such as weather sealing the units, redirecting the pulse away from the probe tips, and addressing static charges from affecting the output. He will give them credit for addressing some of the issues found.

If the probe connection wire has not been addressed, that will still be an issue as it is still a loose connection design, subject to broken contact during movement. This will still cause repeated arc phasing. Whether the internal circuitry monitors the signal pulse before it goes to the output or if it measures from the retuning ground is not known.

Measuring the output before the pulse goes through the cartridge spark gaps is not sufficient. The only method of measuring accuracy is to monitor what pulse is being sent out and comparing that to what the pulse looks like on returning through the ground wire. The spark gap design as covered is not reliable in a system with flexible end point connections. It is still the author's recommendation to remove all the flexible endpoints

1) Modifications recommended by the CEW report.

Several options were considered for improvement modifications;

A redesign to prevent the static electricity from contributing or effecting the output power. Full insulation around the probe and wires is a start. However keeping the probes charged in flight will still lead to seeding the atmosphere. Should this same scenario happen on an insulated carpet, or the same environmental factors, it is unknown if similar results would happen.

After sending these reports to Taser International, they have put out a warning that static electricity can trigger the devices cartridge to fire.

It is unknown if any officer will think of that when firing on a carpeted surface, or any other Triboelectric surface. Static electricity by itself has brought down aircraft before and is unpredictable where, and what concentration it will be. There is not a machine that can give an immediate yes or know answer if it's there. The author's device required some finessing to use.

It is however possible to redesign the product to:
a) Weather seal the unit and probe design.
b) Protect the spark gaps from the environment.
c) Ground the probe wires and redirect static electricity away from the probes and subject before discharging the main pulse.

Recognizing that not every person who is shot by a CEW is a career criminal, or that in past practice when the device was used for less than combative purposes, there should be several power settings available for the officer to choose from. Obviously the device is there as an alternative to firearms, but not every case requires full power.

From it's beginnings as an idea from science fiction to reality, it's intentions of being a non lethal device has not met reality. In science fiction those devices usually have a lethal and non lethal setting, the CEW's only have one, which can do both. A variable or lower setting would be advantageous for individuals who don't need as much convincing, or are not in full muscular fitness health (Or do not resemble the spitting image of physical fitness) to receive a full power setting.

If the skin penetrating design is continued to be used, then some kind of internal resistance measuring method must be done before firing. If the internal resistance of a subject is 25ohms, (as stated by Philips) then a full power charge is not required to be delivered. The Taser International devices are built to overcome 600ohms resistance, a number which they acquired from their animal testing. The difference of delivered output power is 600ohms / 25ohms = 24 times too much power. If they have not already done so, they should also retest all of their devices at the 25 to 180 ohms internal resistance and check again for all side effects.

The principal of electrolysis, when water is exposed to electricity the author does not know how to overcome. That's just a basic science problem that can't be overcome. It will happen whenever electricity comes into contact with water. No one, including police officers, police departments, or the manufacturers could make any guarantee against. Electricity will continue to follow the path of least resistance no matter what the design, the only way to prevent electricity from ever possibly leading the heart, is to prevent it from piercing the only organ designed to stop it from affecting the body, the skin. Without protection from skin, the human body would not be able to survive a walk across a carpet due to the static electricity.

Minor Acidosis can occur from the muscle contraction and some researchers believe this is what is being measured, however Electrolysis causes Acidosis, and the source for the two are indistinguishable from each other. The volume that's created is strictly dependent on the path electricity chooses to take through the body.

2) Taser International discards the testings results

Taser International CEO Rick Smith stated in an CBC / National interview that he dismisses the results of the testing at the Cook County Hospital in Chicago where the CEW testing on pigs showed the heart in a rapid beat during testing (where the chest of the pigs were opened and the heart visually inspected) he said that; "These results would lead to the wrong conclusion, (because the pigs heart are not like a humans heart). There are studies on humans that show this is just not an issue." So this statement effectively dismisses all their animal testing done to date. The dogs were the only other animal which had testing done, so as the dogs heart is also not like a humans heart, the device essentially remains untested for actual medical side effects by the company as of that date.

Taser International has also questioned (trashed) the testing results of the CBC / Radio Canada testing of the police devices. The test was done using the method that the manufacturer recommended to the RCMP, so they have effectively discarded their own testing procedure?

3) Electrical energy can cause tachycardia.

Based on these findings, and the outcome of the Braidwood Inquiry, It is concluded that any electrical charge delivered that reaches the heart, at this point by any CEW, delivering a signal to induce Tetanus and Clonus at more voltage than the action potential of the Cardiac muscle, can and will drive the heart by overriding the Central nervous System, and having it immediately operating at it's maximum potential speed. If the overriding signal is not a continuous wave, a pulsed or intermittent signal can "trigger" the heart at the T stage of the cardiac cycle. If the heart is not conditioned to this rapid or abrupt change, it will lead to the following: An unsynchronized contraction, Ventricular Tachycardia, Cardiac Strain, Ventricular Fibrillation, and / or Cardiac Arrest. The Troponin T marker found will be produced during any of these medical conditions.

The fact that several researchers did find cardiac fibrillation and some did not, seems to indicate that if no other evidence is taken into consideration, may just come down to which particular CEW device was used. The special machine built by Taser International should be independently examined side by side with the other models to compare their outputs to eliminate this device as a source of speculation. This should include design and construction details, maintenance updates and signs of faults or tampering. One researcher reported that the machine is capable of producing fibrillation on the high setting but not on the low setting, so that in itself causes speculation.

This machine may be responsible (even without the testers knowing about it) for many of the published documents relating to the products safety. Battery types and conditions should also be tested to verify that one brand may not be at fault for delivering too much, or not enough power over the other choices. As the author has not seen this machine it is unknown if replicates the cartridge spark gap function, with spark gaps or if it just assumes clean and dry weather operational conditions with an assumed output signal.

If the results in this book are not enough information, then only fair way to decide this he said / she said scenario, in the author's opinion is to gather all the researchers under one roof, and have a side by side comparison of operational regulations, teams, and testing methods.

This should be under an independent panel, consisting of at least medical and veterinary professionals, forensic specialists, electronics specialists, and an independent judge as an overseer. The devices that should be tested (and variants) are the special Taser International testing machine, and any device or model listed in Appendix E, and newer devices on the market since publication. Subjects should be tested for all known side effects because science is based on the ability to repeat the same results under the same conditions, including environmental. The current situation is the manufacturers state the devices are safe because of their testing, and independent researchers and emergency rooms are finding problems. This gathering would be the only fair and quick way to put an end to this science fiction!

4) What other recommendation have been proposed:

a) Multiple discharges are not recommended and should be discouraged. This has been stated by many others before and in the Inquiry.

b) Even when a person is stun only once, the shooter is in no position to determine what internal side effects have taken place, or can assume the health of the individual. Medical treatment should be sought every time.

c) Officers should be trained in the use of portable defibrillators, and CEW equipped officers be required to carry them. This has also been recommended many many times before, including the Inquiry. The officers should have this equipment on hand with the weapon, not sitting back in the squad car, so small portable equipment will have to be developed.

d) Keeping the person in the recovery position (laying on left side) with the head and legs elevated. This should prevent the blood from pooling away, and any VAE (embolisms) from affecting the heart.

e) Training for police to carry and be able to use a Portable Oxygen container is a new recommendation.

Sgt Darren Laur wrote a report for the Canadian Police Research Center in 2004 on Excited Delirium. In it he links the similarities between Excited Delirium and Capture Myopathy normally associated with animals. There are protocols in the Veterinary practice which can be directly applied to human subjects, such as calming measures, cooling measures, intravenous fluids, Vitamin E combined with Selenium, Sodium Bicarbonate to combat Acidosis, and calcium channel blockers. This would be an excellent starting place for research to begin, if it has not already started.

As an alternative, subjects effected fitting this hypothesis or shot by a CEW could be given a sports drink mixture to replenish or correct electrolyte levels, if no better product is available. The author can not find any information if this is being worked on by any company. If not, it appears to be a missed opportunity.

Going further, Ethology is the study of animal behavior processes. The science is not just confined to one animal or species. It can be said that no animal or species likes to be confined, and subjected to pain. They will try to fight back and / or escape every time.

f) Police need to stop pulling their guns to solve issues and adapt methods and tactics from fields other than the military. Just as police have developed the Special Weapons and Tactics (SWAT) to deal with the more extreme elements of crime, the time has come to develop a special team to deal with the mentally unstable or chemically dependent of society. Police forces are stretched everywhere, it is a waste of resources to have officers that should be putting an end to crime, chasing people that need mental or medical help.

g) Skin peircing stun guns should be upgraded to firearm status. In countries that allow civilian ownership, the same requirements as handgun ownership should be considered.

h) Every police TASER® should be verified for correct operation before being issued into service, and subjected to regular verification.

5) An avenue for future research

Veterinary methods use tranquilizer darts to sedate animals that no one in their right mind would approach. This is a direction of research that can be a replacement for CEW's. Veterinary or Animal control officers in North America tranquilize animals as small as badgers to as large as elephants, so why not humans?

Darts are out of the question, as they can be used as a weapon if not seated properly, and can cause injury if shot in the wrong place. What is required is some kind of a product that is fast acting, short duration, and will literally drain the violence out of a person so the officers can restrain them long enough to be placed in custody.

6) What this hypothesis covers.

It is my opinion that by weight of evidence, the primary reason for death by Conducted Electronic Weapons is the replication of the Long QT Syndrome (LQTS). The CEW's by forcefully driving the cardiac muscle at it's maximum potential or maximum intermittent contraction replicates the damages that this syndrome causes. The leading of the heart causes the abnormal re-polarization of the cardiac muscle. If the cardiac muscle is not lead, but pulsed out of sync, this can result in a sprained muscle. A sprained muscle is completely unreliable, and any exertion, or stress will only cause it to falter. The many circumstances and after effects of sudden death can be traced to this event.

The author's reports has explained all described methods of death and side effects that have been recorded and suspected by these devices.
a) The acidosis of the blood has been explained.
b) Medical conditions and side effect conditions have been explained.
c) Additional influences such as drugs & alcohol have been explained.
d) Psychological effects and responses have been described.
e) Differences of the devices and their faults explained.
f) Testing methods, results and manipulation have been explained.

It is the author's strongest recommendation that because of what has been written about QT damage, every person who has been subjected to a CEW discharge, and experienced side effects, that they be medically examined for proper sinus rhythm, especially in the QT area. This condition is difficult to properly diagnose. Individual side effects may vary between persons affected, and non-treatment may only add complications. As of this date the only long term solution to LQTS damage is by an implantable cardioverter-defibrillator (ICD).

It is possible that for every person shot by a CEW, and has LQTS type damage, may require an ICD sometime in their life.

7) What happened to Mr. Dziekanski.

This hypothesis does explain the Dziekanski case. He was agitated, unable to speak the language, throwing computer equipment, had been awake for over 3 days at that point from traveling. Add jet lag, no food, nicotine withdrawal, and no washrooms for the last 10 hours in that part of the airport. His body would have been depleted of any energy reserves, no nourishment, have high level of urea toxins, and be dehydrated.

This author has concluded that based on the results of the postmortem examination and previous research submitted to the inquiry, that the cause of death to Mr. Dziekanski is repeated Arc Phasing of the taser device. High voltage, high current electrical surging due to intermittent contact, (due to the cartridge and probe design) causing leading and / or disruption of the normal sinus rhythm.

This is supported by two events in the Pritchard video:

1) His ability to move / run during the first CEW discharge. This strongly suggests that neuromuscular incapacitation did not work.

2) His vocal yell throughout the video. In Drive Stun (pain compliance mode), subjects are able to yell because the voltage and current are not traveling through the body, but on the outside.

The intermittent connection caused power surging, and lead to extra power being pulsed out by the device. This caused the leading or abnormal sinus rhythm. This causes damage equivalent to Long QT Syndrome. As the electricity jumped between the probe, probe wire and intermittent connection with his body, he endured both partial Neuromuscular Incapacitation and pain compliance mode alternating in the same first 5 seconds. This is the authors explanation for his vocal outbursts modulating and movement.

Just as with any muscle that has been sprained or strained, if a load is suddenly placed on it, it will not react as expected. If a physical load is suddenly placed on the cardiac muscle after an injury, or any visual or emotional stimulation that places a load on the cardiac muscle, will cause it to react as designed, but the injury can affect it any time after that, not just during elevated, rapid or high requirement operation. This can explain sudden deaths within minutes or days after receiving an electrical charge that lead or displaced the cardiac rhythm.

It is the authors opinion that Mr. Dziekanski suffered the Cardiac event at the end of the first discharge, during his fall and spin. This is backed up by evidence on the Pritchard video. When the human body senses it is falling or loosing its balance, it injects a shot of adrenaline. It then uses an involuntary body reaction to rapidly stabilize itself.

Images from the Pritchard video

When the body receives an injury the first instinctive reaction is for your hands to cover the injured area. For example, if you pinch your finger closing a drawer, your non-injured hand will grasp the injured part. If you have a blunt force injury to any part of your body playing a sport, such as hockey, or baseball, you will stop and grab the injured part of your body with your uninjured arm or hand. If multiple areas are injured, then you grab at what ever hurts the most.

What we are seeing in this part of the Pritchard video is Mr. Dziekanski after he falls to the ground, grabbing at his upper chest as he spins around on the ground. Up until the Postmortem was released, the location of the probe impact was not known, so nothing could be guessed at. However, we now know one probe landed in his upper right abdominal muscle, and the other is underneath his right hip, which he is spinning on. The upper probe inserted itself correctly, that is it stuck into the muscle and didn't pull out. The lower probe was only touching his hip, causing arcing and pain. He is not grabbing at either probe when he is spinning on the ground, he is clutching his chest!

The officers didn't testify if he continued to grab his chest when he was facing away from the camera or not. What they did do was one officer had his knee on his throat, one used the CEW in pain compliance mode, and the other two held him down. He was technically already dying from kidney failure when he arrived in the country, however the stun gun triggered some kind of ventricular event. The manufacturer will always argue against this, but the United Nations Committee Against Torture (CAT) has named the stun gun a torture device, in this case, they were right!

From the postmortem standpoint, the lack of urine in his body matches the authors hypothesis of an after effect of Electrolysis (pages 8-11) causing a chemical reaction (graph on page 57). The probes (one seated in the upper right abdominal muscle, which connects to the Falciform ligament) delivering a charge across his lower section, was the most probable route for electricity through his body to start the chemical reaction.

8) Suitability for commercial aircraft use.

After careful consideration, the author must conclude that these devices are not suitable for use in a commercial aircraft because:

a) In a confined environment such as an aircraft, any probes when launched are out of the control of the shooter. Where they actually land on the person is also uncontrollable.

b) The probes don't launch at the same angle so the lower probe is always at risk of injury to another. Even with a laser pointer guiding, the darts are unstabilized projectiles.

c) The device is only good for one shot. (not including the newer models) Multiple hijackers are the current trend.

d) The aircraft is a confined environment, and for the models with three or more cartridges, the wires from the first shot could hinder you from making a second or third by snagging on an object, such as a seat back, armrest or peoples heads.

e) The darts can be easily blocked by readily available objects, such as hostages, or holding objects out in front of you, such as jackets, blankets, pillows, magazines, or staying behind a curtain.

While the aircrews are trained in basic calming measures, and have access to portable Oxygen, medical kits, restraining devices, and depending on the airline or aircraft defibrillators. After a suspect is injured, it falls to the responsibility of that aircrew to provide the medical treatment necessary to keep them alive.

The airlines however don't have a say in what devices the sky marshals use, and the author doesn't have a say in that either, however it would make good sense for the sky marshals to know what equipment is on the aircraft their traveling on, for future use.

There is no recommended weapon for sky marshals to carry, they can carry whatever they want.

What the author recommends

There is always a risk of using a defensive tool or weapon on an aircraft, that the weapon may injure another passenger, crew, or be lost to an attacker, or damage the aircraft and pressurization. When using a tool or weapon, there is a risk that the attacker can gain control of that device, and could be used as another tool to carry out their plans with.

As everyone in airline industry knows what the September 11th hijackers did was to gang up and overpower the defenseless aircrews. There is nothing being done for aircrews since then to stop another gang attack of this type if it were to happen again. The attackers may not have access to small knives or other sharp tools anymore, but anyone can improvise a tool or weapon if necessary.

The flight deck door is not an invincible barrier, and any barrier can be bypassed if not defended properly.

There is a program in the USA where the commercial pilots can carry firearms in the flight deck, an idea that does not sit well with most in the industry, but the idea of arming flight attendants has been rejected because of the already mentioned problems. If both the flight attendants and pilots were armed, there could easily be a cross fire situation, when one person starts shooting, everybody starts shooting!

The only way to stop "bully behavior" is to be able to stand up to it. Therefore in reviewing what's currently left available for aircrews to use, the aircrews must be given training in the only weapon that can't be taken away from them, Martial Arts or Self Defense classes. Flight attendants especially near the flight deck must be able to handle a single attacker, or defend from multiple attackers until the rest of the attendants (or passengers) can gang up on them.

There must be some kind of deterrent mechanism that is publicly shown and displayed, that if you attempt something, we now have the ability to stop you, whereas now there is only the knowledge of the door that is supposedly bullet proof? The benefits are that it does not add unnecessary weight to the aircraft with half measures or window dressings of security, and adds in public confidence of the airline.

Unlike in movies or TV, there is not enough room on an aircraft for any real high kicking maneuvers, however emphasis on Self Defense, Pressure Points, Assailant Defensive Measures in close quarters is appropriate. Training in usage of improvised weapons such as pens, coffee pots, fire extinguishers, oxygen bottles, wine bottles and bottle openers, trolley carts, etc. is also appropriate.

9) The United Nations Committee Against Torture (CAT)

This treaty came into force on June 26th, 1987. The UN has publicly stated that stun guns are a form of torture that can cause death! If the country you live in has signed that treaty, then as of November 2007, any use of that device is in breach of the UN treaty.

The argument is that police officers are enforcing the law. If police officers really are supposedly enforcing the law, then why are they doing it with a device that has been named as a torture device? Each signatory country has signed that they will abide by the rules and regulations set forth within the treaty, so as of November 2007, any country or any person using the stun guns are technically and knowingly violating the UN treaty, however...

Quoting from the treaty;

Having regard to article 5 of the Universal Declaration of Human Rights and article 7 of the International Covenant on Civil and Political Rights, both of which provide that no one shall be subjected to torture or to cruel, inhuman or degrading treatment or punishment,

Article 1
1. For the purposes of this Convention, the term "torture" means any act by which severe pain or suffering, whether physical or mental, is intentionally inflicted on a person for such purposes as obtaining from him or a third person information or a confession, punishing him for an act he or a third person has committed or is suspected of having committed, or intimidating or coercing him or a third person, or for any reason based on discrimination of any kind, when such pain or suffering is inflicted by or at the instigation of or with the consent or acquiescence of a public official or other person acting in an official capacity. It does not include pain or suffering arising only from, inherent in or incidental to lawful sanctions.*

Article 4
1. Each State Party shall ensure that all acts of torture are offenses under its criminal law. The same shall apply to an attempt to commit

torture and to an act by any person which constitutes complicity or participation in torture. 2. Each State Party shall make these offenses punishable by appropriate penalties which take into account their grave nature. (1)

* Inherent in or incidental to lawful sanctions?

This wording has been under much scrutiny and review for years. It was not defined and has left a loophole in the regulations that is being exploited. Author Ahcene Boulesbaa has written a book called "The U.N. Convention on Torture and the prospects for enforcement" which covers the topic in extensive detail. (Martinus Nijhoff Publishers ISBN: 90-411-0457-7, Netherlands).

The UN committee has finally overcome this wording problem by specifically calling the stun gun a torture device! This changes the document as it is now listing the devices which commit torture. Taser International replied 4 days later stating the U.N. Committee Against Torture is "out of touch with the reality that confronts law enforcement officers every day worldwide."

However, as no country has yet signed an amendment to the CAT, its an agreement that is coming into force, but not yet in force. What this now allows for is monitor groups to watch and see who is still buying the stun guns, even though the police, military, (and civilians in some countries) now know its an item that will be banned. The groups can publicly state the dollar amount wasted, and how many devices will need to be thrown away when the ban takes effect.

What you can do;

If you have been shot, injured, or died from a stun gun, your case can be reported to the United Nations, under the section called Petitions. Each signatory country must report their compliance status to the UN every four years, and the committee review's complaints twice per year. *Canada's last report includes information on the stun guns, covering 2004-2007. It was submitted three years late in Oct. 2010? The USA reports are totally centered around Guantanamo Bay.*

Some of the background on the UN statement was the UNCAT was meeting in Portugal at the time and criticized the Portuguese police for recently buying the stun gun, specifically naming the TASER® X26™. An expert had just given the committee a report that the stun guns had "Proven risks of harm or death!"

Amnesty International gave an interview on CBS that "These are people that have seen torture around the world, all kinds of torture. So they don't use the word lightly." (2)

10) What the stun gun has become.

This author's opinion:

The Stun Gun as outlined here has essentially become the so called perfect murder weapon! Police don't want to investigate it. Experts, coroners and doctors don't have the tools to prove the devices can cause cardiac arrest or even give the reasons for acidosis, simply because the reasons are not in any current medical literature.

The manufacturer vigorously defends their products with spokesmen, doctors, nurses, and lawyers. Lawsuits involve everyone, from individuals to inquiries that speak out against, or name their device as a cause of the problem. That list currently includes experts, police, doctors, coroners, other medical staff, engineers, and heads of inquiry! A further side effect is that police departments are willing to sacrifice their own officers before blaming a tool that comes with benefits. Something is obviously wrong! If your an officer in this situation or know of one who is, you too can file a complaint with the UN!

However as stated in the introduction, the intent of this report is not to take away a tool from the police or military. Should an individual consider usage of the devices on anybody that is anything else but the spitting image of physical fitness, in a calm resting or restrained position, should reconsider pulling the trigger.

This page intentionally left blank

Chapter 11 Braidwood Medical Findings

The full Braidwood Inquiry reports can be found online at:

http://www.Braidwoodinquiry.ca

Phase One report is called:

Restoring Public Confidence; Restricting the use of Conducted Energy Weapons in British Columbia.
556 pages.

Phase Two report is called:

Why, The Robert Dziekanski Tragedy. Braidwood Commission on the death of Robert Dziekanski.
470 pages.

Both reports are free .pdf downloads or viewable online in html format. Should that website no longer be available then contact:

http://www.gov.bc.ca/contacts/index.html
EnquiryBC@gov.bc.ca or 1 (604) 660-2421

Excerpts from the medical findings during the inquiry are reprinted with permission.

First, there can be no doubt that an external electrical current can overtake the human body's internal electrical system, resulting in ventricular capture, which may lead to ventricular tachycardia and, in some cases, ventricular fibrillation. Cardiologists routinely introduce small electrical shocks for the purpose of triggering ventricular fibrillation, in order to test newly implanted defibrillators.

Second, there is some evidence that the electrical current from a conducted energy weapon is capable of triggering ventricular capture. The real-life example of a man with a pacemaker having a conducted energy weapon applied against him, resulting in myocardial capture corresponding exactly to the timing of the weapon's pulses, is persuasive evidence that the weapon's electrical current can override the heart.

Third, human studies conducted to date, by researchers such as Drs. Ho, Levine, and Vilke, have not yielded evidence of ventricular tachyarrhythmias. However, I am reluctant to generalize from their studies, for several reasons. They frequently applied the electrical shock to the subject's back as opposed to the chest area, clipped the electrodes to the subject's clothing or taped them to the skin rather than using barbs that penetrate the skin, and in some cases restricted the discharge to 2–3 seconds. More importantly, most of those studies were not capable of ascertaining whether there was an arrhythmia during the weapon's discharge. In one study, the researchers were able to determine that 21 subjects who were monitored by echocardiography had normal heart rhythms during discharge, but that is far too small a sample from which to draw conclusions about whether a weapon is capable of causing ventricular capture and, if so, how frequently.

Fourth, we do know from several animal studies that a conducted energy weapon's discharge can trigger ventricular tachycardia and/or fibrillation in pigs. I approach these studies with caution, recognizing the differences between pigs and humans. Having said that, I am satisfied that it is safe to draw several conclusions from these studies that can be extrapolated to humans. First, the greatest risk of ventricular fibrillation arises when the probes are vectored across the heart. Second, the risk of ventricular fibrillation increases as the tips of the probes get closer to the wall of the heart.

Fifth, I am satisfied that there is a short "window" during the heart's normal beat cycle (the T-wave), when the heart is most vulnerable to an external electrical shock. Fibrillation is known to occur when athletes receive a blow to the sternum during the T-wave, and when cardiologists test newly implanted defibrillators, they time their electrical charge to coincide with the T-wave.

Sixth, while induction of ventricular fibrillation may be dependent on timing of discharges within the vulnerable period of the cardiac cycle, that "narrow window" does not apply to rapid ventricular capture causing ventricular tachycardia, a hemodynamically unstable rhythm which may degenerate into ventricular fibrillation. Death in these circumstances may not be immediate. Ventricular tachycardia is not dependent on timing within the cardiac cycle—discharges at almost any time in the cardiac cycle can capture the heart to cause ventricular tachycardia.

Seventh, while I have concluded that a conducted energy weapon is capable of triggering ventricular capture that may lead to ventricular tachycardia and/or fibrillation, I do not have enough information to quantify that risk with any degree of precision. Further, the risk appears to vary, depending on several factors, which I will discuss later in this part.

Eighth, in deaths proximate to use of a conducted energy weapon, there is often a lack of physical evidence on autopsy to determine whether arrhythmia was the cause of death, which opens the door to debate about whether the weapon or some pre-existing medical condition was responsible. While alcohol or drug intoxication may complicate the pathological analysis in some cases, other explanations must be found in cases where alcohol or drugs were not involved. Several medical experts who made oral presentations during our public hearings emphasized that there must be some explanation for these sudden deaths:

Excited Delirium

On the subject of Excited Delirium, the commission presented the following three Doctors, Lu, Noone, and Vallance on the subject of excited delirium. The following is excerpts from their statements.

a. Dr. Shaohua Lu

Dr. Lu is a psychiatrist at the Vancouver General Hospital's medical surgical unit. He assesses, on a daily basis, patients with severe addictions and severe mental illnesses such as schizophrenia, bipolar disorder, and delirium, and has seven years' previous experience in an emergency psychiatry unit.

Dr. Lu told me that the Diagnostic and Statistical Manual, Volume IV (DSM-IV) defines delirium as a disturbance of the conscious mind, with reduced ability to focus or sustain attention. It involves a change in cognition, or the development of a perceptual disturbance that is not better accounted for by a pre-existing dementia. The disturbance develops over a short time, and tends to fluctuate during the course of the day.

b. Dr. Joseph Noone

Dr. Noone told me that delirium is an acute confusional state with fluctuating levels of consciousness. There is usually hyperactivity, but occasionally lethargy. There is a rapid succession of confused, unconnected ideas, often with illusions (visual misperceptions) and hallucinations. Delirium is most frequently caused by drugs, a closed-head injury, hypoglycemia, electrolyte disturbance, or an acute psychosis such as schizophrenia or manic or bipolar mood disorder.

He emphasized that advanced delirium is a medical emergency, not a psychiatric emergency, requiring intensive medical assessment and management. The goal of treatment is to reverse the causes.

Dr. Noone said that "excited delirium" is not a valid medical or psychiatric diagnosis. In his view it provides a convenient post-mortem explanation for in-custody deaths, where physical and mechanical restraints and conducted energy weapons were employed. His main concern about usage of this term is that "it's being used more and more frequently in an attempt to automatically absolve law enforcement from any and all responsibility for their involvement in sudden in-custody deaths." It would be preferable to:

- avoid use of the term "excited delirium" (which implies a diagnosis), and use the more descriptive, less judgmental term of "emotionally disturbed person"; and

- focus on the principal risk factors for in-custody death—positional asphyxia, cocaine-induced psychosis or delirium, and neuroleptic malignant syndrome.

Dr. Noone said that the best way to treat an emotionally acting-out person is to do the following:

- Assess—you should approach the situation objectively and with an open mind, considering all the possibilities. You should take your time and remain calm. If the degree of force used by the professional is measured on a scale of one to ten, police officers often go in at eight or nine, and this will likely escalate rather than defuse the situation. From his experience, it is much more effective to go in at about three or four—if you go in low, you can usually get compliance.

- Contain—it is unwise for one, two, or even three people to attempt to contain a highly agitated subject. From Dr. Noone's experience, a team response is much more effective. Instead of grabbing the subject roughly, "gentle touching, not touching, showing support is what will bring this confused person down to a level where you can deal with them."

- Treat—once the subject has been transported to the emergency department, it is important to treat the subject for the underlying medical condition. Delirium is a superimposed condition for which there are medical reasons. Often this type of aberrant behaviour is not psychotic—"The big mistake we don't want to make is to treat something as behavioural when in fact it has a medical cause."

When asked about the appropriateness of deploying a conducted energy weapon against a person in a delirious state, Dr. Noone replied: I believe that highly agitated individuals, even more so if they are in delirium, are at very high risk of further medical compromise, due to metabolic, cardiac, respiratory, or other complications.

To "taser" such vulnerable individuals would be contraindicated medically due to the risk of death, in my opinion. That's a clinical opinion.

c. Dr. Maelor Vallance

Dr. Vallance told me that the principal features of delirium are a reduced clarity of awareness of the environment or clouding of consciousness that leads to considerable impairment of attention. Anything coming from the outside, including instructions from the police, may not get through. Even if they do get through, they may not be held in consciousness long enough for the individual to act on them.

There are also changes in cognition, such as disorientation (especially for time), impairments of memory (so that warnings may be quite useless), and problems understanding and expressing language. There may be perceptual disturbances, such as illusions and hallucinations, where they will misinterpret what they hear or see.

Delirium fluctuates. It is usually worse at night because there is less orienting stimulation to keep them on track. It also varies with the level of excitement—the greater the excitement, usually, the greater the derangement.

Dr. Vallance said that there are many underlying causes of delirium, including a general medical condition (e.g., liver disease, AIDS, gross dehydration, electrolyte disturbances or imbalance, acidosis), drug intoxication (especially cocaine and crystal methamphetamines), alcohol withdrawal (e.g., delirium tremens), or a combination of factors (e.g., someone with AIDS who also has cocaine intoxica[tion], someone with pneumonia who has an aberrant reaction [to] medication).

It is not always easy to differentiate delirium from other conditions (e.g., excited catatonic schizophrenia, mania, or agitated dementia), particularly in the community and especially when the person is severely agitated.

Observation alone is insufficient for diagnosis; interactions with the individual and collateral data are also required. Police officers may observe agitation or derangement, but would have no way of knowing the specific condition with which they are dealing.

Dr. Vallance told me that "excited delirium" is not a medical term. It is not described in the medical literature, nor is there real clinical evidence of it as a separate entity. There is no specific pathology post-mortem.

He said:

The symptoms and the behaviours that you see in what is called "excited delirium" are essentially indistinguishable from a deranged, agitated individual in the community, irrespective of the underlying cause. And there is no known specific pathology. In short, there's nothing to identify it.

The first step in intervention should very rarely be physical restraint of any kind. The physical restraint by itself tends to escalate the situation, when the purpose really should be to de-escalate. The escalation of the situation with further agitation and excitement is a danger in itself.

In order to intervene without using physical restraint as a first step, I believe it's necessary to develop specialized intervention teams. There is specific training in that form of intervention. I believe that it's too much to ask the police force to have the level of training that is available as a general training throughout the police. It requires specific selected officers to be specifically trained under our very extensive training programs now. Even then, I don't believe that they should act alone. I believe it should be a team effort.

Chapter 12 Updates April 5, 2011

In 2009 Dr. Mark Debard, heading the taskforce of the The American College of Emergency Physicians formally recognized ExDS (Excited Delirium). (1)

He had the College approve the condition because he saw a video on YouTube and though he could help? David P. Keseg, MD, the medical director of the Columbus Division of Fire, and his colleagues not only describe the condition, but advise the division's emergency medicine personnel on the use of midazolam and ketamine to sedate those with the condition.

EMS personnel in Columbus use the acronym PRIORITY to assess possible excited delirium syndrome patients:

P Psychological issues.
R Recent drug/alcohol use.
I Incoherent thought processes.
O Off (taking clothes off) and sweating.
R Resistant to presence or dialogue.
I Inanimate objects: violent toward shiny or glass objects.
T Tough, unstoppable, superhuman strength.
Y Yelling.

The doctors and College want to distance themselves from the stun gun issue. Any form of force or pain used to stop these people only increases their output, increasing their chances of sudden death. they recommend calming measures. A cocain overdose is almost always the cause of it, followed by drug use and mental instability problems, or combination of medications. These subjects are almost always incapable of understanding or following verbal orders, so officers must retrain themselves from pulling the trigger when their orders are not followed, to another type of capture method.

Generalized tonic-clonic seizure after a taser shot to the head

CMAJ; Esther T. Bui, MD, Myra Sourkes, MD and Richard Wennberg, MD. Mar, 2009

(Tonic-Clonic seizure; Formerly known as Grand Mal Seizures)

The patient was a previously well police officer in his 30s who took part in a police chase involving a suspected robber. He and a colleague cornered the suspect, who initially appeared to surrender but then attempted an escape. The officer had begun to chase the suspect on foot when he experienced a sudden, severe pain in the back of his head. He later described the moment as feeling like he had been "hit by a bat." He recalled letting out a brief gasp before losing consciousness. He had no recollection of falling to the ground on top of the suspect. Police records indicate that the officer's colleague had fired a taser shot meant for the suspect but that the 2 copper darts had instead struck the officer in the occiput and upper back. The officer had been wearing an armoured vest. Immediately after being shot, he was found by his colleague to be unresponsive and foaming at the mouth. His eyes were rolled upward and he had generalized tonic-clonic movements with apnea lasting for about 1 minute. He did not have urinary incontinence. Postictally, he was initially confused and combative. Emergency medical services personnel were able to restrain him. They recorded a Glasgow Coma Score of 9 within 5 minutes after arrival; 5 minutes later, his score was 13.

The patient's next memory was of being in the emergency department. During this period, he felt as if he were in "a bad dream." As he gradually regained orientation over the next few hours, he became aware of thoracic tightness that was aggravated by deep breaths, and a severe headache. He was monitored overnight, then discharged in stable condition.

The patient had no history of febrile or unprovoked seizures, head injuries, headaches, meningitis or encephalitis. He had no family history of seizures or of other neurologic or psychiatric conditions. His developmental history was normal. He was not taking any medications.

The results of a general physical and neurologic examination were normal. Results of routine blood tests were unremarkable except for an elevated leukocyte count of 12.9 (normal 3.6–11.0) x 10⁹/L 30 minutes after the event (decreasing to 11.2 x 10⁹/L 5 hours later) and an elevated serum creatine kinase level of 580 (normal < 232) U/L.

The patient returned to full-time work 5 days after the incident. He experienced persistent headaches, dizziness, back pain and chest tightness. Magnetic resonance imaging scans of the head (1.5 and 3 Tesla) as well as routine and 24-hour ambulatory electroencephalography were performed 1, 2 and 12 months after the seizure. All findings were normal.

A diagnosis of mild traumatic brain injury (concussion), in addition to provoked seizure, was considered after a neurologic consultation during assessment of the patient at a rehabilitation centre 6 months after injury. A psychiatric consultation 7 months after injury suggested an Axis I diagnosis of adjustment disorder with depressed and anxious mood. Formal neuropsychological testing performed 9 months after injury showed no definite evidence of cognitive impairment in any domain.

The patient has not had further seizures since the injury more than 1 year ago. His symptoms of anxiety, difficulties concentrating, irritability, nonspecific dizziness and persistent headaches have not completely resolved. Treatment trials have included amitriptyline 50 mg nightly, topiramate 25 mg nightly, escitalopram 10 mg nightly, almotriptan 12.5 mg as needed and ibuprofen 200–400 mg as needed.

Comments

A taser stun gun is a device designed to temporarily immobilize a human target by delivering a direct-current type of shock through 2 barbed copper darts. The shock causes involuntary muscle contraction. Neuromuscular transmission is thought to be affected primarily at the level of the peripheral motor nerve, although studies have shown that stimulation of the spinal cord may occur with dart penetration as far away as the anterior torso. The muscle contraction induced by tasers is typically tonic, with retained consciousness, no clonic movements and no postictal confusion. The manufacturer's website estimates that a single shot lasts about 5 seconds, delivers 19 pulses per second with a typical charge of 100 microcoulombs per pulse, generates an average net current of 2 milliamperes and has an estimated peak voltage of 1300 volts.

The data are sparse on how this device may affect the central nervous system. A case has been reported involving intracranial penetration by a taser dart with loss of consciousness for 5 minutes. The person who had been struck recovered shortly afterward with a mild headache. No details were reported on whether a seizure occurred, although only 1 of the 2 darts struck the patient. Another case report describes cranial penetration by a taser dart (with the second dart found in a hair braid) with transient decreased consciousness; no further details were given. Other reports of secondary loss of consciousness related to taser shots have involved only cases of severe traumatic head injuries that resulted from falls during neuromuscular incapacitation.

The description by witnesses of the event involving our patient is most compatible with a generalized tonic-clonic seizure. The loss of consciousness, clonic movements, foaming at the mouth and postictal confusion experienced by our patient differentiate the episode from the usual transient incapacitation induced by tasers. The taser current that passed to his brain from the dart in the occiput probably provoked the seizure directly, with a mechanism akin to that of seizures induced by electroconvulsive therapy. In electroconvulsive therapy, an initial charge of 38–60 millicoulombs is used, according to therapeutic protocol in the United States. It is plausible that a copper dart penetrating the scalp and discharging 95 pulses of 100 microcoulombs each could trigger a generalized convulsion.

Given previous case reports of taser-induced cardiac arrhythmias, one could speculate that an initial induced cardiac arrhythmia and a secondary hypoxic seizure, or convulsive syncope, occurred in our case. Convulsive syncope is believed to result from reticular disinhibition in the brainstem resulting from hypoxia-induced cortical dysfunction. However, this mechanism seems unlikely in this case, especially given that the points of impact of the taser darts were over the head and upper back and not the heart.

Even less likely is the possibility that the convulsion was induced by a concussion resulting from the direct physical impact of the darts or impact of the patient's head on the ground. Our patient's prolonged period of unresponsiveness and subsequent postictal confusion is not typical of a concussive convulsion, which is usually characterized by immediate onset and a rapid recovery that takes place over a few minutes. On the other hand, we believe that his persistent symptoms after injury may be attributable in part to postconcussion syndrome, presumably secondary to mild traumatic brain injury caused by either the impact of the taser dart or the subsequent fall to the ground during the provoked seizure.

Until now, most reports of taser-related adverse events have understandably concentrated on cardiac complications associated with shots to the chest. Our report shows that a taser shot to the head may result in brain-specific complications. It also suggests that seizure should be added to the list of taser-related adverse events. (2)

Taser International, which makes the stun gun, says its product warnings already mention the possibility of seizures.

Peter Holran, a Taser spokesman, said the company knew of a few incidents during training in which an officer experienced a seizure after being Tasered. However, in an e-mailed statement, Holran did not specify whether the officers had been hit in the head.

Taser's (new) product warning, posted online, reads: "When practical, avoid intentionally targeting the head, face, throat or genitals without legal justification. Significant injury can occur from device deployment or use into these sensitive areas of the body."

"Repetitive stimuli such as flashing lights or electrical stimuli can induce seizures in some individuals. This risk may be heightened if electrical stimuli or current passes through the head region." (3)

Datrend Systems Verus One; ECD /CEW analyser

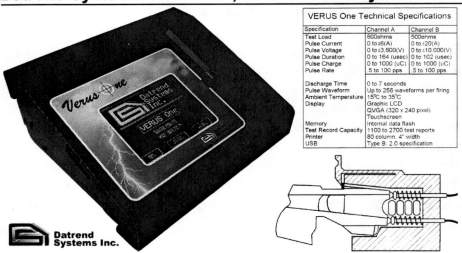

VERUS One Technical Specifications		
Specification	Channel A	Channel B
Test Load	600ohms	500ohms
Pulse Current	0 to ±6(A)	0 to ±20(A)
Pulse Voltage	0 to ±3,600(V)	0 to ±10,000(V)
Pulse Duration	0 to 164 (usec)	0 to 102 (usec)
Pulse Charge	0 to 1000 (uC)	0 to 1000 (uC)
Pulse Rate	.5 to 100 pps	.5 to 100 pps
Discharge Time	0 to 7 seconds	
Pulse Waveform	Up to 256 waveforms per firing	
Ambient Temperature	15°C to 35°C	
Display	Graphic LCD QVGA (320 x 240 pixel) Touchscreen	
Memory	Internal data flash	
Test Record Capacity	1100 to 2700 test reports	
Printer	80 column, 4" width	
USB	Type B, 2.0 specification	

Datrend Systems Inc.

Datrend Systems Inc. is an engineering company that develops, manufacturers, and supports a specialist line of hospital biomedical automated test equipment for Electrical Safety Analysis. They have a line of equipment that specifically tests all aspects of defibrillators, and have used that knowledge to develop the Verus One™, a machine that can currently test the output of the TASER® M26™ and X26™ series. The machine was developed independently to perform the manufacturers recommended field test procedure. The unit is even small enough that it can be taken to the scene of an incident, and the device in question checked on the spot, also giving a paper printout!

As shown in this book, stun guns that test fine in a laboratory setting may test differently out in the environment. The tester only checks the hand held gun itself, not the cartridge, which has to be removed. This is the only product on the market for this job, and the author highly recommends that if your police force or military is currently using the TASER®, you should also invest in this product for obvious reasons.

The machine was developed during the course of Braidwood Inquiry and fulfills the requirements in its conclusions. Should the lower resistance range be adopted later, the analyzer can be modified. Datrend Systems is marketing the Verus One™ directly to the countries police forces and military, so it is not currently advertised on their web site. Contact them directly for more information. www.datrend.com

Construction of the Static Electron Detector

The static electricity detector is based on passed known working designs. It is designed to detect both positive electrostatic voltage and negative electrostatic voltage. This schematic is representative of the prototype when taken in the pictures on 25 December, 2007.

How it works:

The Negative detector will be explained first. The MPF102 transistor is a Junction Field Effect Transistor, and will turn on when the the G pin (connected to an antenna) encounters a positive or negative charge, from the static electricity. The 2N2907 is a PNP transistor, and only when the B pin can connect to ground, through the MPF102, it will activate. The 2N2222 is a NPN transistor. When activated it allows current to flow, but only when the B pin is connected to a positive voltage, through the 2N2907. The Led lights to indicate when the 2N2222 is activated. The 2.2k resistor controls the current flow from the Led, and into the 10milliamp meter. The meters display will increase and decrease as the voltage strength at the gate of the MPF102 increases and decreases.

The Positive detector works the same way except the 2N2907 is replaced with a 2N2222. Only when this transistor receives a positive voltage at the B pin, it will activate. The rest works the same way as the Negative detector. Due to the value of the 2.2k resistor, full scale deflection of the meter is only half way, this was done intentionally to see if the static voltage buildup in the area would saturate the circuit and cause it to read higher than full scale. This did not happen. The metal frame of the enclosure should be connected to ground.

Usage of the meter requires that the antennas be discharged first. When turned on, it will almost always read to full scale deflection, this is normal. This is due to the adhesion factor of electrons. Grounding the antennas alone won't discharge the circuit. Static electricity is everywhere, and the meter is sensitive enough that moving your feet on carpet will keep the sensor fully charged.

The procedure is to repeatedly touch (or pinch) and remove you fingers from the antennas one at a time, until the meters read zero. Then it will be ready for use. Test on a carpet or place in front of a TV.

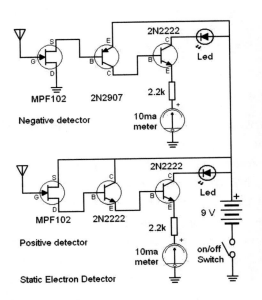

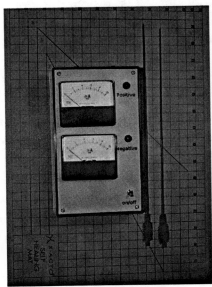

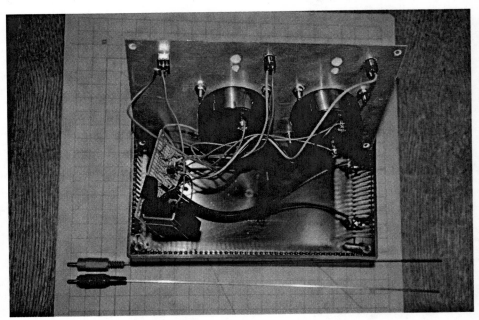

Images and schematic from author.

Note: Each chapter has its own separate reference links.

This information listed below and their links were used between September 07 and March 08.

Chapter 1, CEW report 3

1) Authors hypothesis
2) British Columbia Office of the Police Complaints Commissioner report on "Taser Technology Review final report" June 14, 2005
3) A review of hazards associated with exposure to low voltages, by Dr. M Bikson
4) Carlton University "re: advanced Taser M26 less lethal System"
5) The Darwin awards, Resistance is Futile, 1999
6) Procedural Order Las Vegas Police Department PO 43-04 (taser-lvmpd.pdf)
7) Amnesty International's Report on Tasers, 2004 (amr511392004en.pdf)
8) Pacemakers and Antitachycardia Devices, William Batsford
9) Excited Delirium Syndrome- Cause of death and prevention, Theresa Di Maio, Vincent Di Maio
10) Comparative Analysis of Three Projectile Stun Guns, by Wayne McDaniel
11) Taser Induced Rapid Ventricular Myocardial Capture, Michael Cao, Jerold Shinbane, Jeffrey Gillberg
2) The Advanced Taser M26, X26: Forensic Considerations, by R.T. Wyant
13) Taser Intl press kit 07/19/07
14) RD-SPEC-M26-001_k.pdf
15) RD-SPEC-CRTG-001_J.pdf
16) RD-SPEC_CRTG-003_D.pdf
17) LG-STND-TECDBEC-001 REV C TASER Electrical Characteristics .pdf
18) Stinger Systems provided the following three documents. RBMS Paper 2008.pdf, White Paper Wave Forms-Final.doc, Pig Waveforms 08.pdf. (See updates)

19) Disrupting the heart's tornado in arrhythmia, by Tony Fitzpatrick 2004
20) Balancing the bodies pH, by Lisa Anne Marshall
21) Acid and alkaline, by Herman Aihara
22) Hidden truth of cancer, by Keiichi Morishita
23) Doctors blame Taser stun gun for fibrillation, Alex Bereso 2005
24) Police stun-gun may be lethal firm admits, James Sturcke & Rosalind Ryan 2005
25) The mechanism underlaying sudden death from shock, Geddes LA, Bourland JD, Ford G
26) Dr. Wayne McDaniel Ph.D. University of Missouri
27) Dr. Robert Walter Ph.D.
28) Canadian Broadcasting Corporation, The National www.CBC.ca/national Jan 30/08 et.al.
29) Avionics Fundamentals, Jeppesen Sanderson Training Products, written by United Airlines 1974
30) Excited Delerium Tr-02-2005, by Sgt. Darren Laur.
31) http://www.spectrum.ieee.org/dec07/5731 (How tasers work)
32) http://news-info.wustl.edu/news/page/normal/3750.html
33) http://www.portfolio.mvm.ed.ac.uk/studentwebs/session2/group62/electro.htm
34) http://www.answers.com/topic/defibrillator
35) www.miami-med.com/defibrillators.htm
36) http://www.healthandage.com/html/res/aging_of_you/content/15.html (the hearts pacemaker)
37) http://www.webelements.com/webelements/elements/text/O/key.html
38) http://www.tpub.com/neets/book2/2a.htm
39) http://www.inthesetimes.com/article/2894/ (taser marketing direction)
40) http://van.physics.uiuc.edu/qa/listing.php?id=6793
41) http://hypertextbook.com/facts/1999/CindyAnnRomanowich.shtml
42) http://learn.genetics.utah.edu/units/basics/blood/blood.cfm
43) http://www.officer.com/web/online/Technology/Why-TASERs-Dont-Work/20$38245
44) http://www.crystalradio.net   (inductance calculator)

45) http://www.taser.com/research/Science/Pages/TASERDeviceElectricalDesign.aspx (April 3/08)
46) http://www.att-tactical.com/att_stinger.html
47) http://symtym.com/2004/11/sudden_taser_death_syndrome/
48) http://users.rcn.com/jkimball.ma.ultranet/BiologyPages/M/Muscles.html (tetanus, clonus)
49) http://www.stingersystems.com/
50) http://en.wikipedia.org/wiki/Digoxin
51) http://www.americanheart.org/presenter.jhtml?identifier=64 (human impedance)
52) http://www.cvphysiology.com/Arrhythmias/A002.htm
53) http://www.anaesthesiauk.com/article.aspx?articleid=100573 (defibrillator)
54) http://www.aafp.org/afp/980115ap/groh.html (implantable defibrillator pacemakers)
55) http://truthnottasers.blogspot.com/ (Ongoing death count by CEW devices)
56) http://biz.yahoo.com/pz/080220/136725.html (Department of justice review of the S200)
57) http://www.canada.com/ottawacitizen/news/story.html?id=3fae73fe-af58-4172-8c50-56707c6d47ae&k=8715   (prone position)
58) http://en.wikipedia.org/wiki/Aspiration_pneumonia
59) http://users.rcn.com/jkimball.ma.ultranet/BiologyPages/P/PNS.html#sympathetic (nervous system)
60) http://en.wikipedia.org/wiki/Local_anesthetic
61) http://en.wikipedia.org/wiki/Epinephrine
62) http://en.wikipedia.org/wiki/Catecholamine
63) http://en.wikipedia.org/wiki/Adrenal_medulla
64) http://en.wikipedia.org/wiki/Sedative
65) http://en.wikipedia.org/wiki/Sedation
66) http://en.wikipedia.org/wiki/Anesthesia
67) http://www.aic.cuhk.edu.hk/web8/electrocution.htm
68) http://en.wikipedia.org/wiki/Blood_plasma
69) http://en.wikipedia.org/wiki/Triboelectric_effect
70) http://en.wikipedia.org/wiki/Electrolysis

71) http://en.wikipedia.org/wiki/Ozone
72) http://en.wikipedia.org/wiki/Carbonic_acid
73) http://en.wikipedia.org/wiki/Blood_urea_nitrogen
74) http://en.wikipedia.org/wiki/Electrolyte
75) http://en.wikipedia.org/wiki/Vasopressin (ADH)
76) http://en.wikipedia.org/wiki/Renal_failure
77) http://en.wikipedia.org/wiki/Nitrogen
78) http://en.wikipedia.org/wiki/PH
79) http://en.wikipedia.org/wiki/Haldane_effect
80) http://en.wikipedia.org/wiki/Reperfusion_injury
81) http://en.wikipedia.org/wiki/Cardiac_muscle
82) http://en.wikipedia.org/wiki/Action_potential
83) http://en.wikipedia.org/wiki/Ventricular_fibrillation
84) http://en.wikipedia.org/wiki/Ventricular_tachycardia
85) http://en.wikipedia.org/wiki/Sinoatrial_node
86) http://en.wikipedia.org/wiki/Electrostatic_discharge
87) http://www.tpub.com/neets/book2/2a.htm (self inductance of wire)
88) http://en.wikipedia.org/wiki/Decompression_sickness
89) http://en.wikipedia.org/wiki/Cardiac_arrest
90) http://en.wikipedia.org/wiki/Cocaine
91) http://en.wikipedia.org/wiki/Amphetamine
92) http://en.wikipedia.org/wiki/Alcohol
93) http://en.wikipedia.org/wiki/Nicotine
94) http://www.medical.philips.com/main/products/resuscitation/biphasic_technology/biphasic_intro.html
95) Stinger Systems product brochure S200.specs.pdf (2007)
96) http://en.wikipedia.org/wiki/Ionization
97) http://en.wikipedia.org/wiki/Funny_current
98) Review of Conducted Electronic Weapons, CPRC TR-01-2006
99) http://en.wikipedia.org/wiki/Excited_delirium
100) http://en.wikipedia.org/wiki/Troponin
101) http://en.wikipedia.org/wiki/Voltage-gated_ion_channel
102) http://en.wikipedia.org/wiki/Electric_shock
103) http://www.christopherreeve.org
104) http://en.wikipedia.org/wiki/Electroconvulsive_therapy
105) http://en.wikipedia.org/wiki/Oxygen
106) http://en.wikipedia.org/wiki/Water

107) http://en.wikipedia.org/wiki/Carbon_dioxide
108) http://skydiary.com/gallery/chase2002/2002lightmovie.html
109) The Advanced Taser: A Medical Review, by Anthony Bleetman
110) http://en.wikipedia.org/wiki/Nitrous_oxide
111) http://en.wikipedia.org/wiki/Water_intoxication
112) http://hyperphysics.phy-astr.gsu.edu/Hbase/thermo/electrol.html
113) Advanced Taser X26 Safety Analysis, by The Alfred, Melbourne
114) http://en.wikipedia.org/wiki/Ketamine
115) http://en.wikipedia.org/wiki/Xylazine
116) http://en.wikipedia.org/wiki/Isoflurane
117) http://en.wikipedia.org/wiki/Sevoflurane
118) http://en.wikipedia.org/wiki/Desflurane
119) http://en.wikipedia.org/wiki/Tiletamine
120) http://www.petplace.com/drug-library/telazol-telazol/page1.aspx
121) http://en.wikipedia.org/wiki/Zolazepam
122) http://en.wikipedia.org/wiki/Propofol
123) http://en.wikipedia.org/wiki/Norepinephrine
124) http://en.wikipedia.org/wiki/Dopamine
125) http://en.wikipedia.org/wiki/Neurotransmitter_systems
126) http://en.wikipedia.org/wiki/Sympathetic_nervous_system
127) http://en.wikipedia.org/wiki/Fight_or_flight
128) http://en.wikipedia.org/wiki/Adrenal_gland
129) http://en.wikipedia.org/wiki/Hormone
130) http://en.wikipedia.org/wiki/Phencyclidine
131) http://en.wikipedia.org/wiki/Self-preservation
132) http://en.wikipedia.org/wiki/Human_factors
133) http://en.wikipedia.org/wiki/Ethology
134) http://en.wikipedia.org/wiki/Recovery_position
135) http://www.hc-sc.gc.ca/hl-vs/tobac-tabac/res/news-nouvelles/nicotine_e.html
136) http://whyquit.com/whyquit/LinksAAddiction.html
137) http://www.ncbi.nlm.nih.gov/pubmed/18211316
138) http://www.emedicine.com/emerg/topic312.htm
139) http://en.wikipedia.org/wiki/Metabolic_acidosis
140) Electrical Evaluation of the M-26 Stun Weapon, Final Report, by Shmuel Ben Yaakov
141) http://www.youtube.com/watch?v=GrYkgH5G8Bg (Taser dismisses pig testing, CBC News)

142) http://en.wikipedia.org/wiki/Neuropathy
143) http://www.physorg.com/news64680736.html
144) http://en.wikipedia.org/wiki/Lactic_acid
145) http://www.time-to-run.com/theabc/collapse.htm

References 6,7,23,24,39,50,98 are not specifically quoted but information either for secondary references or supporting documentation was used.

Update 2 April 12, 2008
1) http://au.news.yahoo.com/thewest/a/-/wa/7088886/flat-batteries-can-affect-taser-records/

Update 3 April 21, 2008
1) http://ehealthforum.com/health/topic82737.html

Update 4 August 17 2008

1) Unsolicited report on: Conductive Electronic Weapons; T. Langevin
2) http://www.braidwoodinquiry.ca/
3) UCSD CED Related Studies, by Christian M. Sloane UCSD
4) TASER Protect Life; by Taser International
5) http://en.wikipedia.org/wiki/Short_QT_syndrome
6) http://en.wikipedia.org/wiki/Long_QT_syndrome
7) http://www.mayoclinic.com/health/long-qt-syndrome/DS00434
8) Stinger Systems email to author
9) Cardiac and Physiological Effects of Taser Application: Real World 'implications; by Dr. Zian H. Tseng UCSF
10) http://en.wikipedia.org/wiki/Romano-Ward_syndrome
11) http://en.wikipedia.org/wiki/Implantable_cardioverter-defibrillator

Update 5 January 15 2009

1) http://www.taser.com/research/Science/Documents/XREP.pdf
2) http://allnurses.com/forums/f8/iv-air-bubble-compensation-190642.html
3) http://www.bioengineeringcenter.org/pdfs/The%20Physiological%20Effects%20of%20a%20Conducted%20Electrical%20Weapon%20in%20Swine.pdf
4) http://en.wikipedia.org/wiki/Air_embolism
5) http://emedicine.medscape.com/article/761367-overview
6) http://www.azcentral.com/specials/special43/articles/0521TaserDOD21.html
7) http://www.medical.philips.com/main/products/resuscitation/biphasic_technology/biphasic_intro.html
8) taser_responce_to_cbc.pdf
9) taserinternational_testprotocol.pdf
10) taser-analysis-v1.5 cbc report.pdf
11) http://www.cameronward.com/   (12 Dec 2008)
12) Halifax Chronicle Herald   (17 Dec 2008)
13) The Canadian Press   (17 Dec 2008)
14) http://network.nationalpost.com/np/blogs/posted/archive/2007/12/08/robert-dziekanski-timeline-his-tragic-final-hours.aspx
15) http://incenter.medical.philips.com/default.aspx?tabid=730   (Smart Biphasic Aplication Note Nov 2004.pdf)
16) http://www.infowars.net/articles/may2008/220508Tasers.htm
17) http://infowars.net/articles/may2008/080508Taser.htm
18) http://www.ohio.com/news/18542084.html
19) http://www.nationalpost.com/news/story.html?id=499151
20) http://www.cbc.ca/canada/story/2008/05/07/taser-court.html?ref=rss
21) http://truthnottasers.blogspot.com/

Update 6 April 30 2009

1) http://www.truthnottasers.blogspot.com
2) http://www.yourlawyer.com/articles/read/11338
3) http://www.charlydmiller.com/LIB04/2000petechiaereview.pdf
4) http://www.charlydmiller.com/LIB/1982neckholds.html
5) http://www.charlydmiller.com
6) www.cbc.ca/bc/news/bc-090414-dziekanski-postmortem-exam.pdf
7) http://en.wikipedia.org/wiki/Spark_gap
8) CEW report 3 and updates 1, 2, 3
9) http://en.wikipedia.org/wiki/Long_QT_syndrome
10) http://en.wikipedia.org/wiki/Ventricular_fibrillation

Update 7 July 27, 2009

1) http://en.wikipedia.org/wiki/Lactic_acidosis
2) http://en.wikipedia.org/wiki/Rigor_mortis

Update 8 April 4, 2010

1) http://www.popsci.com/gear-amp-gadgets/article/2009-07/video-taser-tests-new-tri-fire-x3-their-own-employees

Update 9 December 10, 2010
None

Chapter 10
None

Chapter 11
1) http://www2.ohchr.org/english/bodies/cat/index.htm
2) http://www.cbsnews.com/stories/2007/11/25/national/main3537803.shtml

Chapter 12 Updates
1) http://journals.lww.com/em-news/Documents/ExDS-pdf-final.pdf
2) http://www.cmaj.ca/cgi/content/full/180/6/625
3) http://www.vancouversun.com/story_print.html?id=1395638

Appendix C: Braidwood Inquiry List of Presenters

List of phase one information submitters and presenters.

The inquiry stated that over half of the speakers were there because of Taser International. This list is provided for future referencing.

These people made a presentation during the public forums.

*Amnesty International Canada
*BC Association of Municipal Chiefs of Police
*BC Civil Liberties Association
  BC Medical Association
*BC Ministry of Attorney General, Court Services Branch, Sheriff Services

*BC Ministry of Public Safety & Solicitor General, BC Corrections Branch, BC Adult Custody Division

*BC Ministry of Public Safety & Solicitor General, Police Services Division

*BC Ministry of Public Safety & Solicitor General, Police Services Division, Policing and Community Safety Branch

*BC Office of the Police Complaint Commissioner
  BC Schizophrenia Society
  Beattie, Liane
  Beil, Alison
*Bozeman, Dr. William P.
*Butt, Dr. John
*Canadian Mental Health Association, BC Division
*Chambers, Dr. Keith
  Chen, Jinan
*Cisowski, Zofia
*Commission for Public Complaints against the RCMP
  Cook County Trauma Unit
  Creba, Doug
  Crossley, Maynard
  Curry, Ken

Davies, John
Dawson, Robin H.
Dean, Vernan
*Dosanjh, Hon. Ujjal
Excited-Delirium.com (blog owner)
*Gallagher, Cathy & Joseph
Gillman, Patti (Truth Not Tasers.Blogspot.com)
*Greater Vancouver Transportation Authority Police Service
*Hall, Dr. Christine A.
Hantiuk, Margaret
*Ho, Dr. Jeff
Huntley, David
*Janusz, Dr. Michael T.
*Kerr, Dr. Charles R.
Kohne, Horst
*Kosteckyj, Walter
Krzywiecki, Christopher
Langevin, Trevor
Lee, Buddy
*Lu, Dr. Shao-Hua
McDiarmid, Dr. Garnet L.
McDonald, Hunter
McLeod, Lorraine
Moulds, Joy E.
Moyle, Barbarra
*New Westminster Police Department
*Noone, Dr. Joe
Oshanek, Lawrence A.
*Page, Jay
Palys, Ted
*Panescu, Dr. Dorin
Peet, Fred
*Povah, Errol
*Puder, Randy
*Reilly, J. Patrick
*Royal Canadian Mounted Police
Rysstad, Maggie
*Savard, Dr. Pierre

Slater, Kate
Slewidge, Ken
*Sloane, Dr. Christian
Spicer, Phil
*Stethem, Kenneth J.
Street Kid's Project
*Swanson, Jude
*Swerdlow, Dr. Charles
*TASER International, Inc.
*Toronto Police Service
*Tseng, Dr. Zian H.
Ussner, Maryanna
*Vallance, Dr. Maelor
*Vancouver Police Department
*Victoria Police Department
*Ward, Cameron
Watamaniuk, Mark
*Webster, Dr. John G.
*Webster, Dr. Michael
Weitz, Don
White, Natalie

This page intentionally left blank

Appendix D: List of the Dead

The list of the dead as of March 28, 2011. The first 167 were documented by Robert Anglen of the Arizona Republic newspaper. Since December 2005 the list has been recorded by Lawyer Cameron Ward and Patti Gillman. It is not complete and still in progress.

1985

1. April 11 1985, Cornelius Garland Smith, 35, Los Angeles, California

1993

2. March 9 1993, Michael Bryant, 35, Los Angeles, California

1996

3. June 1, 1996: Scott Norberg, 32, Maricopa County, Arizona

1999

4. September 28, 1999: David Flores, 37, Fairfield, California

2000

5. May 14, 2000: Enrique Juarez Ochoa, 34, Bakersfield, California

2001

6. June 17, 2001: Mark Burkett, 18, Gainesville, Florida
7. December 15, 2001: Hannah Rogers-Grippi, 6 months fetus, Chula Vista, California
8. December 17, 2001: Marvin Hendrix, 27, Hamilton, Ohio
9. December 21, 2001: Steven Vasquez, 40, Fort Lauderdale, Florida

2002

10. January 27, 2002: Vincent Delostia, 31, Hollywood, Florida
11. February 12, 2002: Anthony Spencer, 35, Philadelphia, Pennsylvania

12. March 27, 2002: Henry Canady, 46, Hilliard, Florida
13. May 17, 2002: Richard Baralla, 36, Pueblo, Colorado
14. June 10, 2002: Eddie Alvarado, 32, Los Angeles, California
15. June 15, 2002: Jason Nichols, 21, Oklahoma City, Oklahoma
16. June 28, 2002: Clever Craig, 46, Mobile, Alabama
17. June 27, 2002: Fermin Rincon, 24, Fontana, California
18. June 2002: Unidentified male, 39, Phoenix, Arizona
19. July 19, 2002: Johnny Lozoya, Gardena, California
20. July 19, 2002: Gordon Jones, 37, Windermere, Florida
21. September 1, 2002: Frederick Webber, 44, Orange City, Florida
22. November 7, 2002: Stephen Edwards, 59, Shelton, Washington

2003

23. March 16, 2003: Unidentified male, 31, Albuquerque, New Mexico
24. April 16, 2003: Corey Calvin Clark, 33, Amarillo, Texas
25. April 19, 2003: Terrence Hanna, 51, Burnaby, British Columbia
26. May 10, 2003: Joshua Hollander, 22, Normal Heights, California
27. June 9, 2003: Timothy Sleet, 44, Springfield Missouri
28. July 22, 2003: Clayton Willey, 33, Prince George, British Columbia
29. August 4, 2003: Troy Nowell, 51, Amarillo, Texas
30. August 8, 2003: John Thompson, 45, Carrollton Township, Michigan
31. August 17, 2003: Gordon Rauch, 39, Citrus Heights, California
32. September 24, 2003: Ray Austin, 25, Gwinnett, Georgia
33. September 29, 2003: Glenn Leyba, 37, Glendale, Colorado
34. September , 2003: Clark Whitehouse, 34, Whitehorse, Yukon
35. October 7, 2003: Roman Pierson, 40, Brea, California
36. October 11, 2003: Dennis Hammond, 31, Oklahoma City, Oklahoma
37. October 21, 2003: Louis Morris, 50, Orlando, Florida
38. November 6, 2003: James Borden, 47, Monroe County, Indiana
39. November 10, 2003: Michael Johnson, 32, Oklahoma City, Oklahoma
40. November 11, 2003: Kerry O'Brien, 31, Pembroke Pines, Florida
41. December 9, 2003: Curtis Lawson, 40, Unadilla, Georgia
42. December 9, 2003: Lewis King, 39, St. Augustine, Florida

2004

43. February 4, 2004: David Glowczenski, 35, Southampton Village, New York
44. February 12, 2004: Raymond Siegler, 40, Minneapolis, Minnesota
45. February 21, 2004: Curt Rostengale, 44, Silverdale, Washington
46. February 21, 2004: William Lomax, 26, Las Vegas, Nevada
47. March 23, 2004: Perry Ronald, 28, Edmonton, Alberta
48. March 28, 2004: Terry Williams, 45, Madison, Illinois
49. April 1, 2004: Phillip LaBlanc, 36, Los Angeles, California
50. April 16, 2004: Melvin Samuel, 28, Savannah, Georgia
51. April 18, 2004: Alfredo Diaz, 29, Orange County, Florida
52. April 27, 2004: Eric Wolle, 45, Washington Grove, Maryland
53. May 1, 2004: Roman Andreichikov, Vancouver, British Columbia
54. May 13, 2004: Peter Lamonday, 38, London, Ontario
55. May 22, 2004: Henry Lattarulo, 40, Hillsborough, County Florida
56. May 27, 2004: Frederick Williams, 31, Lawrenceville, Georgia
57. May 30, 2004: Darryl Smith, 46, Atlanta, Georgia
58. May 31, 2004: Anthony Oliver, 42, Orlando, Florida
59. June 4, 2004: Jerry Pickens, 55, Bridge City, Louisiana
60. June 9, 2004: James Cobb, 42, St. Paul, Minnesota
61. June 9, 2004: Jacob Lair, 26, Sparks, Nevada
62. June 16, 2004: Abel Ortega Perez, 36, Austin, Texas
63. June 23, 2004: Robert Bagnell, 44, Vancouver, British Columbia
64. June 24, 2004: Kris Lieberman, 32, Bushkill Township, Pennsylvania
65. June, 2004: Bernard Christmas, 36, Dayton, Ohio
66. July 3, 2004: Demetrius Tillman Nelson, 45, Okaloosa County, Florida
67. July 11, 2004: Willie Smith, 48, Auburn, Washington
68. July 17, 2004: Jerry Knight, 29, Mississauga, Ontario
69. July 23, 2004: Milton Salazar, 29, Mesa Arizona
70. August 2, 2004: Keith Tucker, 47, Las Vegas, Nevada
71. August 8, 2004: Samuel Truscott, 43, Kingston, Ontario
72. August 11, 2004: Ernest Blackwell, 29, St. Louis, Missouri
73. August 11, 2004: David Riley, 41, Joplin, Missouri
74. August 13, 2004: Anthony Lee McDonald, 46, Harrisburg, North Carolina

75. August 16, 2004: William Teasley, 31, Anderson, South Carolina
76. August 19, 2004: Richard Karlo, 44, Denver, Colorado
77. August 20, 2004: Michael Sanders, 40, Fresno, California
78. August 24, 2004: Lawrence Davis, 27, Phoenix, Arizona
79. August 27, 2004: Jason Yeagley, 32, Winter Haven, Florida
80. August 29, 2004: Michael Rosa, 38, Del Rey Oaks, California
81. September 12, 2004: Samuel Wakefield, 22, Rio Vista, Texas
82. September 15, 2004: Andrew Washington, 21, Vallejo, California
83. September 20, 2004: Jon Merkle, 40, Miami, Florida
84. October 4, 2004: Dwayne Dunn, 33, Lafayette, Louisiana
85. November 2, 2004: Greshmond Gray, 25, LaGrange, Georgia
86. November 2, 2004: Robert Guerrero, 21, Fort Worth, Texas
87. November 7, 2004: Keith Raymond Drum, Clearlake, Califormnia
88. November 8, 2004: Ricardo Zaragoza, 40, Elk Grove, California
89. November 25, 2004: Charles Keiser, 47, Hartland Township, Michigan
90. November 27, 2004: Byron Black, 39, Lee County, Florida
91. December 4, 2004: Patrick Fleming, 35, Metairie, Louisiana
92. December 15, 2004: Kevin Downing, 36, Hollywood, Florida
93. December 17, 2004: Douglas Meldrum, 37, Wasatch County, Utah
94. December 17, 2004: Lyle Nelson, 35, Columbia, Illinois
95. December 23, 2004: Timothy Bolander, 31, Delray Beach, Florida
96. December 23, 2004: Ronnie Pino, 31, Sacramento, California
97. December 28, 2004: Christopher Hernandez, 19, Naples, Florida
98. December 29, 2004: Jeanne Hamilton, 46, Palmdale, California
99. December 30, 2004: David Cooper, 40, Marion County, Indiana

2005

100. January 2, 2005: Gregory Saulsbury, 30, Pacifica, California
101. January 5, 2005: Dennis Hyde, 30, Akron, Ohio
102. January 7, 2006: Carlos Claros Castro, 28, Davidson City, North Carolina
103. January 8, 2005: Carl Trotter, 33, Pensacola, Florida
104. January 10, 2005, Jerry Moreno, 33, Los Angeles, California
105. January 28, 2005: James Edward Hudson, 33, Chickasha, Oklahoma
106. January 31, 2005: Jeffrey Turner, 41, Lucas County, Ohio

107. February 10, 2005: Ronald Alan Hasse, 54, Chicago, Illinois
108. February 12, 2005: Robert Camba, 45, San Diego, California
109. February 18, 2005: Joel Don Casey, 52, Houston, Texas
110. February 20, 2005: Robert Heston, 40, Salinas, California
111. March 3, 2005: Shirley Andrews, 38, Cincinnati, Ohio
112. March 6, 2005: Willie Towns, 30, Deland, Florida
113. March 12, 2005: Milton Woolfolk, 39, Lake City, Florida
114. March 17, 2005: Mark Young, 25, Indianapolis, Indiana
115. April 3, 2005: James Wathan Jr., 32, Delhi, California
116. April 3, 2005: Eric Hammock, 43, Fort Worth, Texas
117. April 8, 2005: Ricky Barber, 46, Carter County, Oklahoma
118. April 22, 2005: John Cox, 39, Bellport, New York
119. April 24 2005: Jesse Colter, 31, Phoenix, Arizona
120. May 3, 2005: Keith Graff, 24, Phoenix, Arizona
121. May 5, 2005: Kevin Geldart, 34, Moncton, New Brunswick
122. May 6, 2005: Stanley Wilson, 44, Miami, Florida
123. May 6, 2005: Lawrence Berry, 33, Jefferson Parish, Louisiana
124. May 13, 2005: Vernon Young, 31, Union Township, Ohio
125. May 17, 2005: Leroy Pierson, Rancho Cucamonga, California
126. May 20, 2005: Randy Martinez, 40, Albuquerque, New Mexico
127. May 23, 2005: Lee Marvin Kimmel, 38, Reading, Pennsylvania
128. May 23, 2005: Richard Alverado, 38, Tustin, California
129. May 26, 2005: Walter Lamont Seats, 23, Nashville, Tennessee
130. May 28, 2005: Richard T. Holcomb, 18, Akron, Ohio
131. May 28, 2005: Nazario J. Solorio, 38, Escondido, California
132. June 4, 2005: Ravan Conston, 33, Sacramento, California
133. June 7, 2005: Russell Walker, 47, Las Vegas, Nevada
134. June 11, 2005: Horace Owens, 48, Fort Lauderdale, Florida
135. June 13, 2005: Michael Anthony Edwards, 32, Palatka, Florida
136. June 13, 2005: Shawn Pirolozzi, 30, Canton, Ohio
137. June 14, 2005: Robert Earl Williams, 62, Waco, Texas
138. June 24, 2005: Carolyn Daniels, 25, Fort Worth, Texas
139. June 24, 2005: Melinda Kaye Neal, 33, Whitfield County, Georgia
140. June 29, 2005: Pharoah Knight, 33, Miami, Florida
141. June 30, 2005: Gurmeet Sandhu, 41, Surrey, B.C.
142. July 1, 2005: James Foldi, 39, Beamsville, Ont.
143. July 7, 2005: Rocky Brison, 41, Birmingham, Alabama
144. July 12, 2005: Kevin Omas, 17, Euless, Texas

145. July 15, 2005: Ernesto Valdez, 37, Phoenix, Arizona
146. July 15, 2005: Paul Sheldon Saulnier, 42, Digby, Nova Scotia
147. July 15, 2005: Otis G. Thrasher, 42, Butte, Montana
148. July 17, 2005: Michael Leon Critchfield, 40, West Palm Beach, Florida
149. July 18, 2005: Carlos Casillas Fernandez, 31, Santa Rosa, California
150. July 23, 2005: Maury Cunningham, 29, Lancaster, South Carolina
151. July 27, 2005: Terrence L. Thomas, 35, Rockville Centre, New York
152. August 1, 2005: Brian Patrick O'Neal, San Jose, California
153. August 3, 2005: Eric Mahoney, 33, Fremont, California
154. August 4, 2005: Dwayne Zachary, 44, Sacramento, California
155. August 5, 2005: Olsen Ogoddide, 38, Glendale, Arizona
156. August 8, 2005: Unidentified male, 47, Phoenix, Arizona
157. August 26, 2005: Shawn Norman, 40, Laurelville, Ohio
158. August 27, 2005: Brian Lichtenstein, 31, Stuart, Florida
159. September 18, 2005: David Anthony Cross, 44, Santa Cruz, California
160. September 22, 2005: Timothy Michael Torres, 24, Sacramento, California
161. September 24, 2005: Patrick Aaron Lee, 21, Nashville, Tennessee
162. September 26, 2005: Michael Clark, 33, Austin, Texas
163. October 13, 2005: Steven Cunningham, 45, Fort Myers, Florida
164. October 20, 2005: Jose Perez, 33, San Leandro, California
165. October 25, 2005: Timothy Mathis, 35, Loveland, Colorado
166. November 1, 2005: Miguel Serrano, 35, New Britain, Connecticut
167. November 13, 2005: Josh Brown, 23, Lafayette, Louisiana
168. November 17, 2005: Jose Angel Rios, 38, San Jose, California
169. November 20, 2005: Hansel Cunningham, 30, Des Plaines, Illinois
170. November 26, 2005: Tracy Rene Shippy, 35, Fort Meyers, Florida
171. November 30, 2005: Kevin Dewayne Wright, 39, Kelso, Washington
172. December 1, 2005: Jeffrey Earnhardt, 47, Orlando, Florida
173. December 7, 2005: Michael Tolosko, 31, Sonoma, California
174. December 17, 2005: Howard Starr, 32, Florence, South Carolina
175. December 24, 2005: Alesandro Fiacco, 33, Edmonton, Alberta
176. December 29, 2005: David Moss, 26, Omaha, Nebraska

2006

177. January 3, 2006: Roberto Gonzalez, 34, Waukegan, Illinois
178. January 5, 2006: Matthew Dunlevy, 25, Laguna Beach, California
179. January 13, 2006: Daryl Dwayne Kelley, 29, Houston, Texas
180. January 16, 2006: Shmekia Lewis (female), 24, Beaumont, Texas
181. January 22, 2006: Nick Ryan Hanson, 24, Ashland, Oregon
182. January 25, 2006: Murray Bush, Metairie, Louisiana
183. January 27, 2006: Jorge Luis Trujillo, San Jose, California
184. January 28, 2006: Karl W. Marshall, 32, Kansas City, Missouri
185. January 29, 2006 Benites Sichero, 39, Spokane County, Washington
186. January 31, 2006: Jaime Coronel, Castroville, California
187. February 6, 2006: Jessie Williams Jr., 40, Harrison County, Mississippi
188. February 13, 2006: Darval Smith, New Orleans, Louisiana
189. February 19, 2006: Gary Bartley, 36, Mandeville, Louisiana
190. February 24, 2006: Samuel Hair, 48, Fort Pierce, Florida
191. March 4, 2006: Melvin Anthony Jordan, 27, Norman, Oklahoma
192. March 8, 2006: Robert R. Hamilton, 42, St. Augustine, Florida
193. March 18, 2006: Otto Zehm, 35, Spokane, Washington
194. March 18, 2006, Cedric Davis, 26, Merced County, California
195. March 20, 2006: Timothy Grant, 46, Portland, Oregon
196. March 24, 2006: Theodore Rosenberry, 35, Hagerstown, Maryland
197. April 5, 2006: Thomas Clint Tipton, 34, Clearwater, Florida
198. April 15, 2006: Nick Mamino Jr., 41, St. Louis, Missouri
199. April 16, 2006: Billy Ray Cook, 39, Dublin, North Carolina
200. April 16, 2006: Juan Manuel Nunez III, 27, Lubbock, Texas
201. April 18, 2006: Richard McKinnon, 52, Cumberland County, North Carolina
202. April 21, 2006: Alvin Itula, 35, Salt Lake City, Utah
203. April 24, 2006: Jose Romero, 23, Dallas, Texas
204. April 24, 2006: Emily Marie Delafield, 56, Green Cove Springs, Florida
205. May 1, 2006: Jeremy Davis, 24, Bellmead, Texas
206. May 7, 2006: Kenneth Cleveland, 63, Ashtabula, Ohio
207. May 25, 2006: Brian Davis, 43, Los Angeles, California
208. June 4, 2006: Felipe Herrera, 48, Las Vegas, Nevada

209. June 13, 2006: Jerry Preyer, 43, Pensacola, Florida
210. June 18, 2006: Jason Troy Dockery, 31, Coolville, Tennessee
211. June 21, 2006: Kenneth Eagleton, 43, Crosby, Texas
212. June 21, 2006: Joseph Stockdale, 26, Indianapolis, Indiana
213. June 24, 2006: John Martinez, San Jose, California
214. July 2, 2006: Jermail Williams, 32, South Bend, Indiana
215. July 7, 2006: Michael Deon Babers, 26, Shreveport, Louisiana
216. July 8, 2006: Christopher Tull, 36, Cincinnati, Ohio
217. July 9, 2006: Nickolos Cyrus, 29, Mukwonago, Wisconsin
218. July 11, 2006: Jesus Negron, 29, New Britain, Connecticut
219. July 20, 2006: Mark McCullaugh, 28, Akron, Ohio
220. July 23, 2006: Shannon Johnson, 37, Pittsboro, North Carolina
221. August 2, 2006: Anthony Jones, 39, Merced, California
222. August 4, 2006: Ryan Michael Wilson, 22, Lafayette, Colorado
223. August 8, 2006: Curry McCrimmon, 26, Melbourne, Florida
224. August 8, 2006: James Nunez, 27, Santa Ana, California
225. August 9, 2006: Glen Thomas, 33, Wabasso, Florida
226. August 17, 2006: Raul Gallegos-Reyes, 34, Centennial, Colorado
227. August 21, 2006: Timothy Picard, 41, Woonsocket, Rhode Island
228. August 23, 2006: Noah Lopez,25, Fort Worth, Texas
229. August 30, 2006: Jason Doan, 28, Red Deer, Alberta
230. September 1, 2006: Juan Soto, Jr., 39, Liberal, Kansas
231. September 4, 2006, Jesus Mejia, 33, Los Angeles, California
232. September 5, 2006: Larry Noles, 52, Louisville, Kentucky
233. September 8, 2006: Perry Simmons, 35, Montgomery, Alabama
234. September 13, 2006: Laborian Simmons, 24, Marion County, Florida
235. September 17, 2006: Marcus Roach-Burris, 42, Menasha, Wisconsin
236. September 17, 2006: James Philip Chasse Jr., 42, Portland, Oregon
237. September 29, 2006: Joseph Kinney, 36, Madison Twp., Ohio
238. September 30, 2006: Vardan Kasilyan, 29, Las Vegas, Nevada
239. September 30, 2006: John David Johnson III, 27, Orange Park, Florida
240. October 1, 2006: Kip Darrell Black, 38, North Charleston, South Carolina
241. October 5, 2006: Michael Templeton, 50, Jonesboro, Arkansas

242. October 6, 2006: Herman Carroll, 31, Houston, Texas
243. October 9, 2006: James Simons, 35, Lincoln Park, Michigan
244. October 19, 2006: James Lewis, 37, Las Vegas, Nevada
245. October 19, 2006: Nicholas Brown, Milford, Connecticut
246. October 22, 2006: Jordan Case, 20, Tualatin, Oregon
247. October 22, 2006: Eddie Charles Ham Jr., 30, Montgomery, Alabama
248. October 23, 2006: Michael Todd Gleim, 40, Milford, Ohio
249. October 29, 2006: Roger Holyfield, 17, Jerseyville, Illinois
250. October 30, 2006: Jeremy Foos, 29, Columbus, Ohio
251. November 9, 2006: William Jobe, 40, Federal Way, Washington
252. November 14, 2006: Timothy Wayne Newton, 43, Rocky Mount, North Carolina
253. November 14, 2006: Darren Faulkner, 41, Southaven, Mississippi
254. December 3, 2006: Briant K. Parks, 39, Columbus, Ohio
255. December 17, 2006: Terrill Enard, 29, Lafayette, Louisiana
256. December 30, 2006: Daniel Walter Quick, 43, Magalia, California

2007

257. January 5, 2007: Calvin Thompson, 42, Gastonia, North Carolina
258. January 6, 2007: Douglas John Ilten, 45, Fort Pierce, Florida
259. January 7, 2007: Blondel Lassegue, 38, Queen's, New York
260. January 9, 2007: Pete Carlos Madrid, 44, Fresno, California
261. January 17, 2007: Keith Kallstrom, 56, Milan, Michigan
262. January 18, 2007: Andrew J. Athetis, 18, Gilbert, Arizona
263. January 29, 2007: Michael Keohan, 45, Huntingdon Park, California
264. January 30, 2007: Christopher L. McCargo, 43, Dayton, Ohio
265. February 11, 2007: Stephen Krohn, 44, Mesa, Arizona
266. February 21, 2007: Martin Mendoza, 43, Oceanside, California
267. March 13, 2007: Muszack Nazaire, 24, East Naples, Florida
268. March 15, 2007: Randy Buckey, 42, Marion, Ohio
269. March 16, 2007: Ryan Lee Myers, 40, Essex, Maryland
270. March 17, 2007: David Brown, 47, Park Forest, Illinois
271. March 17, 2007: Unidentified male, West Covina, CA
272. March 23, 2007: Sergio Galvan, 35, San Antonio, Texas
273. April 10, 2007: Eugene Donjuall Gilliam, 22, Prattville, Alabama

274. April 11, 2007: Roberto Perez, 25, Indio, California
275. April 14, 2007: Unidentified male, Phoenix, Arizona
276. April 22, 2007: David Mills, 26, Hamden, Connecticut
277. April 23, 2007: Unidentified male, Houston, Texas
278. April 24, 2007: Louis Jermaine Broomfield, 35, Charleston, South Carolina
279. April 24, 2007: Walter Heller, 55, Santa Rosa, California
280. April 24, 2007: Uywanda Peterson, 43, Baltimore, Maryland
281. April 30, 2007: Roy Hamner, 59, Pearl, Mississippi
282. May 5, 2007: Daniel Bradley Young, 33, Seminole, Florida
283. May 7, 2007: Robert A. Keske, 45, Seminole, Florida
284. May 12, 2007: Trent A. Yohe, 37, Spokane, Washington
285. May 12, 2007: Jeffry Young, 54, Bremerton, Washington
286. May 14, 2007: Terrill Heath, 31, Baltimore, Maryland
287. May 15, 2007: Chance W. Shrum, 20 years old, Iola, Kansas
288. May 16, 2007: Patrick D. Hagans, 42, Valleyview, Ohio
289. May 19, 2007: Milisha Thompson, 35, Oklahoma City, Oklahoma
290. May 22, 2007: Kevin DeWayne Hill, 39, Knoxville, Tennessee
291. May 23, 2007: Raymundo Guerrerro Garcia, 33, Simi Valley, California
292. May 25, 2007: Steve Salinas, 47, San Jose, California
293. May 26, 2007: Marcus D. Skinner, 22, Seat Pleasant, Maryland
294. May 29, 2007: Doyle Moniki Jackson, 34, Benton Harbor, Indiana
295. May 29, 2007: Ramel Henderson, 51, San Diego, California
296. June 19, 2007: Juan Flores Lopez, 47, San Angelo, Texas
297. July 2, 2007: Richard Baisner, 36, Pasadena, California
298. July 8, 2007: Nathaniel Cobbs Jr., 25, Newburgh, New York
299. July 16, 2007: Albert Romero, 47, Denver, Colorado
300. July 20, 2007: Jermaine Thompson, 36, Kansas City, Missouri
301. July 25, 2007: Carlos Rodriguez, 27, Atlanta, Georgia
302. July 29, 2007: Ronald Marquez, 49, Phoenix, Arizona
303. August 2, 2007: Clyde Patrick, 44, Birmingham, Alabama
304. August 4, 2007: Gefery Johnston, 42, Chicago, Illinois
305. August 4, 2007: Stephen Spears, 49, Detroit, Michigan
306. August 4, 2007: James Barnes, 21, Omaha, Nebraska
307. August 11, 2007: Craig Berdine, 37, Fremont, Ohio
308. August 14, 2007: Rafael Forbes, 21, Jackson, Mississippi
309. August 15, 2007: James Wells, 43, Waterford, California (no. 275)

310. August 18, 2007: Thomas Campbell, 50, Baltimore, Maryland
311. August 23, 2007: Chad Cekas, 27, Pittsburgh, Pennsylvania
312. August 26, 2007: Glenn Shipman Jr., 44, Portland, Oregon
313. September 3, 2007: Earl Guerrant, 47, Golf Manor, Ohio
314. September 3, 2007: Charles Gordon, 26, Vallejo, California
315. September 9, 2007: Jorge Renteria Terrquiz, 25, Anaheim, California
316. September 20, 2007: Claudio Castagnetta, 32, Quebec City, Quebec
317. October 1, 2007: Samuel Baker, 59, Quitman, Georgia
318. October 1, 2007: Keith White, 44, Kansas City, Kansas
319. October 12, 2007: Michael Patrick Lass, 28, Orange County, California
320. October 14, 2007: Robert Dziekanski, 40, Richmond, BC
321. October 14, 2007: Donald Clark Grant, 54, Asheville, North Carolina
322. October 17, 2007: Quilem Registre, 39, Montreal, Quebec
323. November 1, 2007: Seldon Deshotels, 56, Lake Charles, Louisiana
324. November 2, 2007: Stefan McMinn, 44, Hendersonville, North Carolina
325. November 7, 2007: Roger Brown, 40, Miami, Florida
326. November 16, 2007: Paul Carlock, 57, Springfield, Illinois
327. November 18, 2007: Jesse Saenz, 20, Raton, New Mexico
328. November 18, 2007: Jarrel Gray, 20, Frederick, Maryland
329. November 18, 2007: Christian Allen, 21, Springfield, Florida
330. November 20, 2007: Conrad Lowman, Jacksonville, Florida
331. November 22, 2007: Howard Hyde, 45, Halifax, Nova Scotia
332. November 24, 2007: Robert Knipstrom, 36, Chilliwack, British Columbia
333. November 29, 2007: Ashley R. Stephens, 28, Ocala, Florida
334. November 30, 2007: Cesar Silva, 32, Los Angeles, California
335. December 10, 2007: Leroy Patterson Jr., 41, Walton County, Georgia

2008

336. January 2, 2008: Brandon Smiley, 27, Mobile, Alabama

337. January 5, 2008: Ryan Rich, 33, Las Vegas, Nevada
338. January 9, 2008: Otis C. Anderson, 36, Fayetteville, North Carolina
339. January 11, 2008: Xavier Jones, 29, Coral Gables, Florida
340. January 15, 2008: Mark Backlund, 29, New Brighton, Minnesota
341. January 17, 2008: Baron Pikes, 21, Winnfield, Louisiana
342. January 18, 2008: Daniel Hanrahan, 44, Staten Island, New York
343. February 3, 2008: Louis Cryer, 32, Port Arthur, Texas
344. February 3, 2008: Joseph Davis, 50, Brandon, Mississippi
345. February 7, 2008: Richard Earl Abston, 53, Merced, California
346. February 19, 2008: Garrett Sean Farn, 41, Bakersfield, California
347. February 26, 2008: Barron Harvey Davis, 44, Mayes County, Oklahoma
348. March 4, 2008: Christopher Jackson, 37, Clay, New York
349. March 6, 2008: Javier Aguilar, 46, Roswell, New Mexica
350. March 18, 2008: Roberto Gonzalez, 24, Chicago, Illinois
351. March 20, 2008: Darryl Wayne Turner, 17, Charlotte, North Carolina
352. March 21, 2008: James Garland, 41, Deerfield Beach, Florida
353. March 29, 2008: Henry Bryant, 35, Indianapolis, Indiana
354. March 30, 2008: Walter Edward Haake Jr., 59, Topeka, Kansas
355. April 2, 2008: Jason Jesus Gomez, 35, Santa Ana, California
356. April 6, 2008: Yvelt Occean, 31, New Kent County, Virginia
357. April 22, 2008: Uriah Samson Dach, 26, Richmond, California
358. April 24, 2008: Kevin Piskura, 24, Cincinnati, Ohio
359. April 24, 2008: Dewayne Chatt, 39, Memphis, Tennessee
360. April 27, 2008: Paul Thompson, 24, Greensboro, North Carolina
361. April 28, 2008: Jermaine Ward, 28, Jackson, Tennessee
362. May 4, 2008: Joe Kubat, 21, St. Paul, Minnesota
363. May 6, 2008: James S. Wilson, 22, Alton, Missouri
364. May 28, 2008: Ricardo Manuel Abrahams, 44, Woodland, California
365. May 31, 2008: Robert Ingram, 27, Raceland, Louisiana
366. June 5, 2008: Willie Maye, 43, Birmingham, Alabama
367. June 6, 2008: Donovan Graham, 39, Meriden, Connecticut
368. June 8, 2008: Quintrell T. Brannon, 25, Vincennes, Indiana
369. June 9, 2008: Tony Curtis Bradway, 26, Brooklyn, New York
370. June 23, 2008: Jeffrey Marreel, 36, Norfolk, Ontario

371. June 24, 2008: Ernest Graves, 26, Rockford, Illinois
372. June 27, 2008: Nicholas Cody, 27, Dothan, Alabama
373. July 2, 2008: Isaac Bass, 34, Louisville, Kentucky
374. July 4, 2008: Othello Pierre, 23, Baton Rouge, Louisiana
375. July 8, 2008: Samuel DeBoise, 29, St. Louis, Missouri
376. July 8, 2008: Carlos Vargas, 42, San Bernardino, California
377. July 14, 2008: Marion Wilson Jr., 52, Houston, Texas
378. July 14, 2008: Deshoun Keyon Torrence, 18, Long Beach, California
379. July 22, 2008: Michael Langan, 17, Winnipeg, Manitoba
380. July 23, 2008: Richard Smith, 46, Dallas, Texas
381. July 26, 2008: Anthony Davidson, Statesville, 29, North Carolina
382. August 4, 2008: Jerry Jones, 45, Beaumont, Texas
383. August 4, 2008: Andre Thomas, 37, Swissvale, Pennsylvania
384. August 2, 2008: Lawrence Rosenthal, 54, Hemet, California
385. August 10, 2008: Kiethedric Hines, 31, Rockford, Illinois
386. August 15, 2008: Kenneth Oliver, 45, Miami, Florida
387. August 25, 2008: Ronald Adkisson, 59, Creston, Iowa
388. August 29, 2008: Stanley James Harlan, 23, Moberly, Missouri
389. September 3, 2008: Prince Swayzer, 38, San Jose, California
390. September 3, 2008: Andy Tran, 32, Garden Grove, California
391. September 11, 2008: Roney Wilson, 46, Hillsborough, Florida
392. September 17, 2008: Sean Reilly, 42, Mississauga, Ontario
393. September 19, 2008: Gabriel Bitterman, 23, Lincoln, Nebraska
394. September 25, 2008: Iman Morales, 35, New York, New York
395. September 30, 2008: Frank Frachette, 49, Langley, BC
396. October 1, 2008: Jose Anibal Amaro, 45, Orange County, Florida
397. October 18, 2008: Homer Taylor, 39, Chicago, Illinois
398. October 29, 2008: Trevor Grimolfson, 38, Edmonton, Alberta
399. October 31, 2008: Marlon Oliver Acevedo, 35, Riverside, California
400. November 2, 2008: Gordon Walker Bowe, 30, Calgary, Alberta
401. November 3, 2008: Adren Maurice Turner, 44, Mexia, Texas
402. November 10, 2008: Guy James Fernandez, 42, Santa Rosa, California
403. December 3, 2008: Leroy Hughes, 52, Covington, Kentucky
404. December 9, 2008: Quincy Smith, 24, Minneapolis, Minnesota
405. December 19, 2008: Edwin Rodriguez, 26, San Jose, California

406. December 21, 2008: Nathan Vaughn, 39, Santa Rosa, California
407. December 24, 2008: Mark Green, 46, Houston, Texas

2009

408. January 8, 2009: Derrick Jones, 17, Martinsville, Virginia
409. January 11, 2009: Rodolfo Lepe, 31, Bakersfield, California
410. January 22, 2009: Roger Redden, 52, Soddy Daisy, Tennessee
411. February 2, 2009: Garrett Jones, 45, Stockton, California
412. February 11, 2009: Richard Lua, 28, San Jose, California
413. February 13, 2009: Rudolph Byrd, Age Unknown, Thomas County, Florida
414. February 13, 2009: Michael Jones, 43, Iberia, Louisiana
415. February 14, 2009: Chenard Kierre Winfield, 32, Los Angeles, California
416. February 28, 2009: Robert Lee Welch, 40, Conroe, Texas
417. March 22, 2009: Brett Elder, 15, Bay City, Michigan
418. March 26, 2009: Marcus D. Moore, 40, Freeport, Illinois
419. April 1, 2009: John J. Meier Jr., 48, Tamarac, Florida
420. April 6, 2009: Ricardo Varela, 41, Fresno, California
421. April 10, 2009: Robert Mitchell, 16, Detroit, Michigan
422. April 13, 2009: Craig Prescott, 38, Modesto, California
423. April 16, 2009: Gary A. Decker, 50, Tuscon, Arizona
424. April 18, 2009: Michael Jacobs Jr., 24, Fort Worth, Texas
425. April 30, 2009: Kevin LaDay, 35, Lumberton, Texas
426. May 4, 2009: Gilbert Tafoya, 53, Holbrook, Arizona
427. May 6, 2009: Grant William Prentice, 40, Brooks, Alberta
428. May 17, 2009: Jamaal Ray Valentine, 27, La Marque, Texas
429. May 23, 2009: Gregory Rold, 37, Salem, Oregon
430. June 9, 2009: Brian Layton Cardall, 32, Hurricane, Utah
431. June 13, 2009: Dwight Jerome Madison, 48, Baltimore, Maryland
432. June 20, 2009: Derek Kairney, 36, South Windsor, Connecticut
433. June 29, 2009: Shawn Iinuma, 37, Fontana, California
434. July 2, 2009: Rory McKenzie, 25, Bakersfield, California
435. July 30, 2009: Jonathan Michael Nelson, 27, Riverside County, California
436. August 9, 2009: Terrace Clifton Smith, 52, Moreno Valley, California

437. August 12, 2009: Ernest Owen Ridlehuber III, 53, Greenwood, South Carolina
438. August 14, 2009: Hakim Jackson, 31, Philadelphia, Pennsylvania
439. August 18, 2009: Ronald Eugene Cobbs, 38, Greensboro, North Carolina
440. August 20, 2009: Francisco P. Sesate, 36, Mesa, Arizona
441. August 22, 2009: T.J. Nance, 37, Arizona City, Arizona
442. August 26, 2009: Miguel Molina, 27, Los Angeles, California
443. August 27, 2009: Manuel Dante Dent, 27, Modesto, California
444. September 7, 2009: Shane Ledbetter, 38, Aurora, Colorado
445. September 16, 2009: Alton Warren Ham, 45, Modesto, California
446. September 19, 2009: Yuceff W. Young II, Brooklyn, Ohio
447. September 21, 2009: Richard Battistata, 44, Laredo, Texas
448. September 28, 2009: Derrick Humbert, 38, Bradenton, Florida
449. October 2, 2009: Rickey R. Massey, 38, Panama City, Florida
450. October 12, 2009: Christopher John Belknap, 36, Ukiah, California
451. October 17, 2009: Frank Cleo Sutphin, 19, San Bernardino, California
452. October 27, 2009: Jeffrey C. Woodward, 33, Gallatin, Tennessee
453. November 13, 2009: Herman George Knabe, 58, Corpus Christi, Texas
454. November 14, 2009: Darryl Bain, 43, Long Island, New York
455. November 16, 2009: Matthew Bolick, 30, East Grand Rapids, Michigan
456. November 17, 2009: Edward Buckner, 53, Chattanooga, Tennessee
457. November 19, 2009: Jesus Gillard, 61, Bloomfield Township, Michigan
458. November 21, 2009: Ronald Petruney, 49, Washington County, Pennsylvania
459. December 10, 2009: Hatchel Pate Adams III, 36, Hampton, Virginia
460. December 11, 2009: Paul Martin Martinez Jr., 36, Roseville, California
461. December 11, 2009: Andrew Grande, Panama City Beach, 23, Florida
462. December 13, 2009: Douglas Boucher, 39, Mason, Ohio
463. December 20, 2009: Preston Bussey III, 41, Rockledge, Florida

464. December 21, 2009: Michael D. Hawkins, 39, Springfield, Missouri
465. December 30, 2009: Stephen Palmer, 47, Stamford, Connecticut

2010

466. January 6, 2010: Delano R. Smith, 21, Elkhart, Indiana
467. January 17, 2010: William R. Bumbrey III, 36, Arlington, Virginia
468. January 20, 2010: Kelly Brinson, 45, Cincinnati, Ohio
469. January 27, 2010: Joe Nathan Spruill Jr., 33, Goldsboro, North Carolina
470. January 28, 2010: Patrick Burns, 50, Sangamon County, Illinois
471. January 28, 2010: Daniel Mingo, 25, Mobile, Alabama
472. February 8, 2010: Mark Andrew Morse, 36, Phoenix, Arizona
473. March 4, 2010: Roberto Olivo, 33, Tulare, California
474. March 5, 2010: Christopher A. Wright, 48, Seattle, Washington
475. March 10, 2010: Jaesun Ingles, 31, Midlothian, Illinois
476. March 10, 2010: James J. Healy, 44, Rhinebeck, New York
477. March 19, 2010: Albert Valencia, 31, Downey, California
478. April 10, 2010: Daniel Joseph Barga, 24, Cornelius, Oregon
479. April 30, 2010: Adil Jouamai, 32, Arlington, Virginia
480. May 9, 2010: Audreacus Davis, 29, DeKalb County, Georgia
481. May 14, 2010: Sukeba Olawunmi, 39, Clarkston (Dekalb County), Georgia
482. May 24, 2010: Efrain Carrion, 35, Middletown, Connecticut
483. May 28, 2010: Carl D'Andre Johnson, 48, Baltimore, Maryland
484. May 29, 2010: Anastasio Hernandez, 42, San Ysidro (San Diego), California
485. May 29, 2010: Jose Martinez, 53, Waukegan, Illinois
486. June 9, 2010: Terrelle Leray Houston, 22, Hempstead, Texas
487. June 13, 2010: William Owens, 17, Homewood, Alabama
488. June 14, 2010: Jose Alfredo Jimenez, 42, Harris County, Texas
489. June 15, 2010: Michael White, 47, Vallejo, California
490. June 22, 2010: Daniel Sylvester, 35, Crescent City, California
491. June 24, 2010: Aron Firman, 27, Collingwood, Ontario
492. July 5, 2010: Damon Lamont Falls, 31, Oklahoma City, Oklahoma
493. July 5, 2010: Edmund Gutierrez, 22, Imperial, California
494. July 8, 2010: Phyllis Owens, 87, Boring, Oregon
495. July 9, 2010: Marvin Louis Booker, 56, Denver, Colorado

496. July 12, 2010: Anibal Rosario-Rodriguez, 61, New Britain, Connecticut
497. July 18, 2010: Edward G. Stephenson, 46, Leavenworth, Kansas
498. July 23, 2010: Jermaine Williams, 30, Cleveland, Mississippi
499. August 1, 2010: Dennis C. Sandras, 49, Houma, Louisiana
500. August 9, 2010: Andrew Torres, 39, Greenville, South Carolina
501. August 18, 2010: Martin Harrison, 50, Dublin, California
502. August 19, 2010: Adam Disalvo, 30, Daytona Beach, Florida
503. August 20, 2010: Stanley Jackson, 31, Superior Township, Michigan
504. August 23, 2010: Michael Ford, 50, Livonia, Michigan
505. August 25, 2010: Eduardo Lopez-Hernandez, 21, Las Vegas, Nevada
506. August 31, 2010: King Ramses PJG Hoover, 27, Spanaway, Washington
507. September 4, 2010: Adam Collier, 25, Gold Bar, Washington
508. September 4, 2010: Adam Collier, 25, Gold Bar, Washington
509. September 10, 2010: Larry Rubio, 20, Lemoore, California
510. September 12, 2010: Freddie Lee Lockett, 30, Dallas, Texas
511. September 16, 2010: Gary Lee Grossenbacher, 48, Oklahoma City, Oklahoma
512. September 17, 2010: David Cornelius Smith, 28, Minneapolis, Minnesota
513. September 18, 2010: Joseph Frank Kennedy, 48, La Mirada, California
514. October 4, 2010: Javon Rakestrau, 28, Lafayette, Louisiana
515. October 7, 2010: Patrick Johnson, 18, Philadelphia, Pennsylvania
516. October 10, 2010: Michael Bain, 31, Billings, Montana
517. October 14, 2010: Karreem A. Ali, 65, Montgomery County, Maryland
518. October 20, 2010: Troy Hooftallen, 36, Gaskill Township, Pennsylvania
519. November 4, 2010: Eugene Lamott Allen, 40, Wilmington, Delaware
520. November 4, 2010: Mark D. Shaver, 32, Kent, Ohio
521. November 6, 2010: Robert A. Neill Jr., 61, Mount Joy, Pennsylvania

522. November 23, 2010: Denevious Thomas, 36, Doughterty County, Georgia
523. November 25, 2010: Rodney Green, 36, Waco, Texas
524. November 27, 2010: Blaine Terrell McElroy, 37, Jackson County, Mississippi
525. December 2, 2010: Clayton Early James, age unknown, Elizabeth City, North Carolina
526. December 11, 2010: Anthony Jones, 44, Las Vegas, Nevada
527. December 13, 2010: Linel Lormeus, 26, Naples, Florida
528. December 21, 2010: Christopher Knight, 35, Brunswick, Georgia
529. December 31, 2010: Rodney Brown, 40, Cleveland, Ohio

2011

530. January 5, 2011: Kelly Wayne Sinclair, 41, Amarillo, Texas
531. February 5, 2011: Robert Ricks, 23, Alexandria, Louisiana
532. February 24, 2011: Unidentified male, age unknown, Los Angeles, California
533. March 14, 2011: Christopher Davis, 36, Los Angeles, California
534. March 15, 2011: Brandon Bethea, 24, Harnett County, North Carolina
535. March 21, 2011: Jerry Perea, 38, Albuquerque, New Mexico

Appendix E: Comparison of Devices · 227

This list only includes the stun guns with skin piercing probes.

TASER International, USA

Air TASER™ or M34000™

Voltage	Freq.	Cartridge	Cost	Multishot
		Obsolete		

M26™ (aka M18L™ Obsolete)

Voltage	Freq.	Cartridge	Cost	Multishot
50kv	50kHz 20pps,15pps	15,25,35ft	$500.	No

M26C™

Voltage	Freq.	Cartridge	Cost	Multishot
50kv	50kHz 20pps,15pps	15ft	$500.	No

X26™

Voltage	Freq.	Cartridge	Cost	Multishot
50kv	100kHz 19pps	15,25,35ft	$1000.	No

X26C™

Voltage	Freq.	Cartridge	Cost	Multishot
50kv	100kHz 17pps,10pps	15ft	$1000.	No

C2™

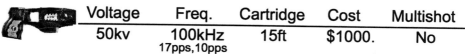

Voltage	Freq.	Cartridge	Cost	Multishot
50kv	?Hz 19pps,8pps	15ft	$350.	No

X3™ aka WILDLIFE TASER™

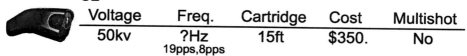

Voltage	Freq.	Cartridge	Cost	Multishot
50kv	?Hz 15pps ea	15,25,35ft	$1800.	Yes 3

XREF™ (Requires 12 gauge shotgun)

Voltage	Freq.	Range	Cost	Multishot
450-550v	?Hz 19pps	65ft	$360.	No

Shockwave™

Voltage	Freq.	Cartridge	Cost	Multishot
50kv	?Hz	25ft	$1700./6	Yes Tiered array

Stun Wire Shot Technology©, Russia
(New products, not yet on market)

Pull Down Gun S5

Voltage	Freq.	Cartridge	Cost	Multishot
75kv	?Hz	10M,15M	?	Yes 5

Legionary-T (also uses Taser International cartridges)

Voltage	Freq.	Cartridge	Cost	Multishot
?	?Hz	10M,15M	?	Yes 3+1

SWS (Multishot Adapter for X26™)

Voltage	Freq.	Cartridge	Cost	Multishot
N/A	N/A	10M	?	Yes 4

Leyden Gun

Voltage	Freq.	Cartridge	Cost	Multishot
?	?Hz	10M,15M	?	Yes 30

MART, Russia

Air 107U (Several models available with different lengths)

Voltage	Freq.	Cartridge	Cost	Multishot
120kv	260Hz	15ft	?	No

AIR 107U-S Karakurt

Voltage	Freq.	Cartridge	Cost	Multishot
120kv	260Hz	15ft	$500.RU	No

L3 Communications, USA

Sticky Shocker™ 37mm (Requires SL-6, M203, or M79)

Voltage	Freq.	Range	Cost	Multishot
50kv	50Hz 15pps,10pps	10M	?	No

Phazzer Electronics, USA

Cartridges compatible with Taser International M26™, X26™

Phazzer Dragon™

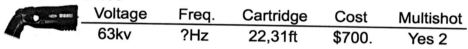

Voltage	Freq.	Cartridge	Cost	Multishot
80kv	?Hz 15pps	15ft	$329.99	No

Phazzer Enforcer™

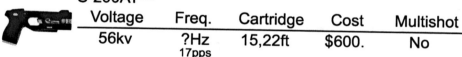

Voltage	Freq.	Cartridge	Cost	Multishot
75kv	?Hz 15pps	15ft	$599.99	No

Stinger Systems, USA

Classed as a firearm, (Out of business, see page 158)

S-400™

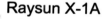

Voltage	Freq.	Cartridge	Cost	Multishot
63kv	?Hz	22,31ft	$700.	Yes 2

S-200AT™

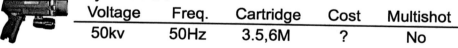

Voltage	Freq.	Cartridge	Cost	Multishot
56kv	?Hz 17pps	15,22ft	$600.	No

Jiun-An Technology, Taiwan

Classed as a firearm. Uses solid propellant to deploy probes

Raysun X-1A

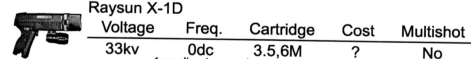

Voltage	Freq.	Cartridge	Cost	Multishot
50kv	50Hz	3.5,6M	?	No

Raysun X-1D

Voltage	Freq.	Cartridge	Cost	Multishot
33kv	0dc .1ma direct current	3.5,6M	?	No

All information and images are taken from the manufacturers web sites or product documentation. Not all information was publicly available.

General certification and safety specifications.

Russia is the only country with government regulations concerning the stun guns since 1996. Document GOST 50940-96 covers the operation, manufacture, sale, safety, marketing and warranty of all stun gun products that are manufactured there.

Of the requirements covered, it states that:

- No device shall cause injury to the user.
- Will not cause biological or pathological effects in the body.
- All side effects should not last more than 20 to 30 minutes.
- Safe to use in all weather.
- No more than 25% lost power through 2 1/2 inches of heavy clothing.
- Devices are built for three categories, through heavy winter clothing and all weather usage, medium weight clothing, and summer clothing.
(The devices built for summer weight clothing have reduced power output requirements than devices built for heavy winter clothing.
- Spacing of probes defined.
- Power levels and waveforms are defined.
- All devices maintain the same electrical specification for 1000 applications.
- All devices have a 12 month warranty.

The rest of the world leaves it to the manufacturers to assume all and come up with whatever specifications they desire.

The GOST_50940-96 document (Printed in Russian) is hosted at:
http://stun-wire-shot-technology.narod.ru/GOST_50940-96.pdf